ENGINEERING TEST PRINCIPLES
FOR
OPERATIONAL SUITABILITY

Special attention has been given to laying out the information in the book. Thus, closely related information is presented on facing pages wherever possible to minimize the amount of page turning that might otherwise be required. Setting the book out in this manner resulted in a number of blank pages being interspersed through the text. These pages can be used for additional notes.

ENGINEERING TEST PRINCIPLES FOR OPERATIONAL SUITABILITY

Edited by

Elizabeth Rodriguez
OFFICE OF THE SECRETARY OF DEFENSE
Office of Director of Defense,
Research and Engineering

A. DEEPAK Publishing **1992**
A Division of Science and Technology Corporation
Hampton, Virginia USA

A. DEEPAK Publishing
A Division of Science and Technology Corporation
101 Research Drive
Hampton, Virginia 23666–1340   USA

Library of Congress Cataloging-in-Publication Data

    Engineering test principles for operational suitability
    Elizabeth Rodriguez.
        p. cm.
    Includes bibliographical references.
    ISBN 0-937194–23–9 (hardcover) : $32.00
    1. United States—Armed Forces—Equipment-
supplies—Testing. I. Rodriguez, Elizabeth.
UC263.E52   1992
355.8'028'7—dc20

Printed in the United States of America

Table of Contents

Preface

This book is an updated version of the two-volume Operational Suitability Guide written for the test and evaluation of defense weapon systems. Edited by Dr. Elizabeth Rodriguez, ODDDR&E(P&R), this book combines the two volumes into a single document. Volume I of the Operational Suitability Guide has been incorporated into Chapters 1, 2, 3, and 8, and Volume II has been incorporated into Chapters 4, 5, 6, and 7.

Special thanks are due to the many contributors for their participation, including Mr. James Duff, OPTEVFOR; Col. David Woodruff and Mr. Dave Young, AFOTEC; Mr. Fred McCoy and Mr. Gary Bates, OTEA; Lt. Col. Chuck Walters and Maj. Don Creighton, MCOTEA; Col. Ted Cress, ODDR&E(R&AT); and Mr. Marty Meth and Mr. Thomas J. Parry, OASD(P&L)WSIG, who provided guidance and advice in the development of the Guide, and extensive personal time in reviewing drafts and coordinating and collecting comments with their respective services and agencies. Mr. Thomas A. Musson and Mr. John R. Rivoire were instrumental in the analyses and background research that were fundamental to the writing of the manuscript.

Chapter 1

INTRODUCTION

1.1 PURPOSE

The purpose of this book is to assist test engineers in their preparation of and/or review of OT&E documents and in observing OT&E events. It highlights important operational suitability factors for consideration during these activities. The book is structured to aid in the preparation of and/or the review and evaluation of the three principal operational test and evaluation documents: Test and Evaluation Master Plans (TEMPs), OT&E plans, and OT&E reports.

1.2 DEFINITIONS

Operational Test and Evaluation is conducted to determine the operational effectiveness and suitability of weapons, equipment, or munitions for use in combat by typical military users. The Department of Defense defines operational suitability and operational effectiveness in the following manner:

Operational Suitability: "The degree to which a system can be placed satisfactorily in field use with consideration given to availability, compatibility, transportability, interoperability, reliability, wartime usage rates, maintainability, safety, human factors, manpower supportability, logistics supportability, documentation, and training requirements."

Operational Effectiveness: "The overall degree of mission accomplishment of a system when used by representative personnel in the environment planned or expected for operational employment of the system considering organization, doctrine, tactics, survivability, vulnerability, and threat (including countermeasures, nuclear, and chemical and/or biological threats)."

Operational effectiveness and operational suitability, in a strict sense, cannot be separated. There are elements within each that easily could be included in the other.

1.3 OVERVIEW

This book is structured around the four key activities of the operational test engineer: the three documents which must be prepared and/or reviewed (TEMPs, plans, and reports) and the observation of actual testing, with the intent of focusing on specific suitability issues that will be reported in the review of each. The book is organized as follows:

Chapter 2 contains a tutorial on the 13 elements of operational suitability.

Chapter 3 contains a tutorial on other operational suitability issues.

Chapter 4 contains guidance for the review of TEMPs.

Chapter 5 contains information for conducting the review of OT&E Plans.

Chapter 6 contains information for observing the conduct of Operational Testing.

Chapter 7 contains information for conducting the review of the OT&E Reports.

Chapter 8 contains a copy of Annex A ("Common Reliability, Availability and Maintainability Terms for use in Multi-Service OT&E Test Programs.") to the 1989 OTAs Memorandum of Agreement.

The subsections of Chapters 4 through 7 consist of templates that are subdivided into the following sections:

- The introduction presents an overview of the subject of the template, e.g., what the particular part of the OT&E document should contain.

- The area of risk identifies the risk that might be encountered if proper test planning, conduct, or reporting of suitability concerns is not followed.

- The Outline for reducing risk directs the Staff Assistant to what might be included in a proper response to the subject of the template. Each of the items under the outline for reducing risk is further emphasized by an example (enclosed in a box) of its application to a DoD system.

1.4 HOW TO USE THIS BOOK

An understanding of operational suitability and the suitability elements is a critical component of suitability OT&E. The relationship of these elements to each other and to the successful introduction of the system into the operating forces is an important and necessary part of systems acquisition.

The reader may wish to review the 13 elements of suitability, what parameters are appropriate for measuring those elements, and how thresholds for those elements might be checked. Such a review can be aided by reading Chapters 2, 3, and 8.

The reader can review and evaluate the complexity of operational suitability by briefly scanning the templates in Chapters 4, 5, 6, and 7. However, to take full advantage of the guidance offered, the reader should study this book from front to back, thus gaining a mental road map for quick access to reference material when it is needed.

For a specific document under review, guidance on TEMPS is provided in Chapter 4; on OT&E plans, Chapter 5; or on test reports, Chapter 7.

Information relative to observing operational testing is provided in Chapter 6.

Chapter 2

OPERATIONAL SUITABILITY

INTRODUCTION

Operational suitability is defined as

the degree to which a system can be placed satisfactorily in field use with consideration given to availability, compatibility, transportability, interoperability, reliability, wartime usage rates, maintainability, safety, human factors, manpower supportability, logistics supportability, documentation, and training requirements.

The suitability elements listed in this definition are discussed on the following pages. However, the definition lists only some of the items that may be suitability issues for a particular system; additional issues may be dictated by the system's mission(s) or the planned logistics support process that, during operational test and evaluation (OT&E) planning, may need to be considered. Chapter 3 discusses five of these additional issues: suitability modeling and simulation, integrated diagnostics, environmental factors, electromagnetic environmental effects (E3), and software suitability.

Chapter 2 and 3 are organized to present, as sub-sections, a definition of each suitability element or issue to be discussed, some of the parameters that apply, and key points that must be considered in its application to operational suitability. Key points and milestone activities that serve to support the overall operational suitability aspect of OT&E are set forth below.

KEY POINTS

Effective systems must be suitable.

Poor suitability may preclude the use of an otherwise effective system in combat. Any limitation that suitability imposes on the effective employment of a weapon system must be identified and evaluated. The suitability portion of OT&E must be planned and conducted to determine what, if any, limitation is likely to exist when the system is placed in operation.

Suitability issues that have the highest risk must be identified.

While all suitability issues must be satisfactory, the risks associated with these issues vary. The array of suitability topics for an individual system usually involves a number of suitability critical operational issues (COIs). Identifying these critical issues allows focus and attention to be directed to those areas in need of detailed and careful examination. Developing this focus is an essential part of the early Service planning of every OT&E. Independent examination of the system description, mission description, and planned support concept by the DOT&E staff provides the basis for evaluating Service-identified critical suitability issues.

KEY POINTS (Cont'd)

The operating scenario drives the suitability demands.

The demands for maintenance and supply support depend on the intensity of the system's use. Operating hours per day, miles per day, flying hours per day, that portion of the operation conducted at high speed conditions, any ratio of different mission types, etc., all are important factors for estimating the intensity of operational use. These factors must be estimated for the planned operational scenario and then considered as part of the Services' planning for the operational testing. The Services specify these mission parameters and scenarios in various documents, e.g., Operational Mode Summary. The way in which the system is used during operational testing, including the realism of the mission scenario(s) and the environmental conditions, is critical to constructing a test program that provides realistic operational suitability demonstrations and produces realistic suitability test data.

Terminology needs to be consistent.

Table 2-1 presents a hierarchy of terminology that relates the conduct of OT&E, i.e., characteristic, parameter, threshold. The evaluation of operational suitability involves selecting and examining *characteristics* that relate to each of the suitability issues. The characteristics for each of the issues are selected, considering the system's operating and support scenarios and the relative criticality of the areas included in each of the issues. As part of the planning for operational test, each characteristic will have one or more *parameters* identified. To evaluate the data that result from the operational tests in each of these areas, the parameter measurements must be compared to *thresholds* that were previously established. The table includes an example of how these terms might be applied when reliability is an operational suitability issue.

Table 2-1 Hierarchy of Terminology

TERM	EXAMPLE
Characteristic	Mission Reliability
Parameter	Mean Time Between Operational Mission Failures (MTBOMF)
Threshold	MTBOMF Shall Be at Least 300 Hours

There are always limitations to operational testing.

Resource and safety constraints often impose limitations on the conduct of the testing. The number of test articles available, the number of test hours, the availability of all support systems, and the realism of the logistics system almost always are limited. The importance of these limitations and their effect on the suitability results must be addressed in the Test and Evaluation Master Plan (TEMP), test plan, and test report.

Operational suitability applies to each level of support.

The Services use various support structures (i.e., levels of support) for weapons systems. Table 2-2 presents some examples of the levels of support that may apply. For most systems, the suitability elements (e.g., maintainability, training, documentation) apply differently to each support level. While only the first and second levels of support may be available at the operational test site, the evaluation should consider, if possible, all applicable support levels.

Table 2-2 Variance in the Definitions of Support Levels

Level	Type of Support	Example		
		A	B	C
1st	Owner or User	Organizational	Crew	Crew
			Unit	Unit
2nd	Supporting Unit(s) With More Capability	Intermediate	Direct Support	Direct Support
				General Support
3rd	Highest Level of Capability	Depot	Depot	Depot

Operational suitability has many dimensions.

It is impossible to combine the many quantitative and qualitative aspects of suitability into a single measure -- the unique aspects of each system impose a different priority on the suitability issues. The overall assessment of a system's suitability is an expert judgment based upon a multitude of factors. The suitability evaluation must assess what the OT results say about the likelihood that the system can be satisfactorily placed in field use in the intended operating environment. If an area of suitability is less than the level stated in the requirements, the evaluation should estimate the impact of this deficiency on the system.

MILESTONE ACTIVITIES

During the early system definition studies and analyses, the critical operational issues (COIs) in the suitability area should be identified. The initial Test and Evaluation Master Plan (TEMP) should discuss these issues. The characteristics that relate to COIs should be identified by Milestone I, the Concept Demonstration/Validation Decision. The system mission profile(s) and life profile also should be defined by Milestone I and documented in the Operational Mode Summary, or similar document. (The life profile is a time-phased description of the events and environments that an item experiences from manufacture to final expenditure of the item or its removal from the operational inventory. It includes one or more mission profiles, in addition to any storage, transportation, maintenance, or exercise events and environments that the item will experience.) The early definition of these profiles does not imply that the profiles are "in concrete" and will not be revised.

By Milestone II, the Full-Scale Development Decision, the program manager and the developing contractors should have a reasonably well defined "system-level" design. The required level of reliability and maintainability should be known. The maintenance diagnostics approach, the maintenance concept, and the general level of support requirements should be established. The training concept should be understood. The relationship between the system's reliability and maintainability requirements and the maintenance concept should be defined and in balance with the planned logistics support concept. High reliability systems can be supported with unique logistics support systems, e.g., missiles that are handled as "wooden rounds." (A "wooden round" is a missile or munition that is handled in the operating unit as a single assembly. There is no plan or capability to isolate faults or to disassemble the item at the operating unit.) Weaknesses in these areas or lack of detailed knowledge may cause problems as the program proceeds. Lack of definition at Milestone II also may result in the developing organizations or contractors having differing views of some aspects of the program. This can lead to inconsistency between the support planning and the system's detailed design. The definition of the system and its support concept are needed to define the OT&E criteria and produce the Milestone II TEMP.

If there is a Milestone IIIA (the Low Rate Initial Production (LRIP) Decision), an operational assessment report or the first of the OT&E reports should show the status of the system in meeting its operational suitability requirements and satisfactorily resolving the suitability issues. Testing should be complete in some operational suitability areas, and results compared to the criteria and the evaluation results reported during the Defense Acquisition Board (DAB) process.

The OT&E report that supports Milestone IIIB (the Full-Rate Production Decision) should update the previous information and, if not complete, should have an expanded evaluation of operational suitability subjects. The test and evaluation report should compare the results with the threshold, highlight the current "status" of the system, and describe areas that have changed status, i.e., from "deficient" or "unsatisfactory" to "satisfactory." This report should contain the final assessment of the question, "Is the system suitable?"

2.1 AVAILABILITY

Availability is defined as

a measure of the degree to which an item is in an operable and committable state at the start of a mission when the mission is called for at an unknown (random) time.

This definition of availability addresses systems that spend a portion of their time in a "ready" status, and at some undetermined time are required to initiate a mission. The discussion of a system's availability must consider the type of system being considered. Items being operationally tested range from entire aircraft, to complex ship combat systems, to relatively small man-portable systems and items that are only part of a complex combat system. Some systems spend most of their time in a readiness status, always available to perform a single mission (e.g., a strategic missile system). Other types of systems with other operating scenarios require other measures. Some are "continuous-use" systems that are required to perform twenty-four hours a day (e.g., command and control computer systems, communications systems, or warning systems). Other systems spend time in a ready status and perform repetitive missions when called upon (e.g., tactical aircraft). In many cases, the call to perform a mission is not necessarily random; the operational commander exercises some control over when the particular system is required to perform a mission. Because of the degree of control over the scheduling of tactical aircraft sorties, some aircraft systems are better characterized by using "sorties per aircraft per day," or "sortie rate," as a measure of how available the system is to perform its mission. The probability of being available for the first mission may be very different than the probability of being available for second and later missions. Examination of the availability for each of these cases requires consideration of the different perspective in each case. In some cases, manpower levels may limit availability.

PARAMETERS

The multi-Service memorandum of agreement (MOA) (Chapter 8) has two definitions for operational availability, A_o. The first is

$$A_o = \frac{\text{Total Uptime}}{\text{Total Uptime} + \text{Total Downtime}}$$

when operated in an operational mission scenario. The second is

$$A_o = \frac{\text{Number of systems ready}}{\text{Number of systems possessed}}$$

The first equation can be used during the OT of subsystems, and in situations where it is possible for the system to be in a state other than "up" or "down." An example would be a situation where there is an interruption of the testing for redeployment of the test forces, system reconfiguration, or other activity. The operational test plan must state clearly how these periods of "no test" are defined and who will determine when these periods start and stop.

The Services use other methods of calculating the ratio of availability. For some Army and Marine Corps systems, the A_O is calculated by an expanded equation

$$A_o = \frac{\text{Operating Time + Standby Time}}{\substack{\text{Operating Time + Standby Time + Total Corrective Maint. Time} \\ \text{+ Total Preventive Maint. Time + Total Administrative and} \\ \text{Logistics Downtime}}}$$

At other times, the Services may use parameters such as the percent of time that the system is Mission-Capable (MC), Full Mission-Capable (FMC), and Partial Mission-Capable (PMC) as measures for availability. These measures are dependent on the list of system equipment that is essential for each system's missions. (A full listing of mission-essential equipment should be contained in the TEMP, or included by reference.) By definition,

> a system is "Full Mission-Capable" when it has all mission-essential equipment available and can perform any of its missions;

> a system is "Partial Mission-Capable" when only a portion of the mission-essential items are available, but can perform at least one, but not all, of its missions;

> a system is "Mission-Capable" when it is in either a PMC or FMC condition;

> "Not Mission-Capable" means that the system does not have the equipment available to perform any of its missions.

One of the advantages of these parameters is that they may allow OT&E results to be put into terms that are familiar to operational commanders.

Another availability measure is achieved availability. This parameter may be used in situations where the test is limited in the logistics area. It is very costly to procure spares or to have a representative logistics system for engineering development models (EDM) or for situations where two or more contractors are competing by producing competitive systems that are to be evaluated in an operational test. When the supply support is limited and nonrepresentative, the total administrative and logistic downtime component of operational availability cannot be evaluated. Achieved availability does not consider the downtime associated with logistic or administrative delays. The equation for achieved availability is

$$\frac{\text{Achieved}}{\text{Availability}} = \frac{\text{Operating Time}}{\substack{\text{Operating Time + Total Corrective Maintenance Time} \\ \text{+ Total Preventive Maintenance Time}}}$$

Another parameter for availability relates to the percentage of time that an item is able to satisfy a demand for its service. This parameter may be applied to systems that are drawn from storage or from a stockpile, or to systems that are maintained in a standby state and then called on to support a mission at a specific time. Demand availability is usually expressed as a percentage.

$$\text{Demand Availability} = \frac{\text{Number of times available}}{\text{Number of times requested}}$$

Additional parameters for availability are listed in the multi-Service MOA (See Chapter 8).

KEY POINTS

Availability is a critical characteristic that should be discussed in the early planning documents.

The early system planning documents should provide a basis for relating availability to other system characteristics as the program proceeds through later acquisition phases. The system requirements should identify which availability parameter is most meaningful for the system. How this parameter relates to the operating scenario and the reliability and maintainability parameters needs to be understood.

System availability is difficult to measure during short operational testing periods.

During operational testing, the measured value for availability can be totally unrepresentative of what might be expected in operational service. For example, in a short test period, only a few failures may occur. As this may not be a representative number of failures, the resulting calculated availability may be very optimistic. For an immature system, the time to identify problems and restore the system to an operational status may be extremely lengthy. In these cases, the limitations on the availability measure must be recognized. In addition, the planned maintenance and supply systems may not be in place because the system is not yet fielded and may not be fielded. Modeling and simulation may be useful in assessing the availability in these situations.

The OT planning should address the methods of measuring times for the availability evaluation.

The way in which a system is to be used in operation will determine the most appropriate availability parameter. How that parameter is applied to the system in question will help establish how the system's operational time is recorded and evaluated. If the system has standby time, or time that will not be included in the availability calculation, then the test planning should address the specifics of how these items will be handled. If the OT is to include periods during which the measure of time is not to be included in the availability evaluation, then the test plan should address the definition and likelihood of "no test" time, and indicate how this time will be addressed in calculating operational availability.

System standby time may be important.

If system standby time is included in the calculation of availability, then the ratio of the standby time to active system operating time should be assessed for "reasonableness." If an unreasonably high ratio of standby time to system operating time is evident in the test, then the calculated operational availability will be unrealistically high. An estimate of the ratio that will be observed in combat operations should be described in the Operational Mode Summary, or similar documents. The planning for the operational testing should address what ratio is planned for the testing period, and how this compares with the estimated ratio for operational use.

Logistics support realism should be an objective in planning for operational testing.

For useful insight into the operational availability of the system being tested, the logistics support being provided during the testing should be as realistic as possible. Any limitations that exist need to be identified prior to the test, and included in the evaluation of test results.

Mode transitions also should be evaluated.

If a failure occurs immediately following standby time, i.e., when the system transitions to an active mode, then there is a question about the "true" condition of the system while it was in standby. In short-duration operational tests, an occurrence such as this can have a significant impact on calculated availability. A conservative estimate might be made by assuming that the system was down during the entire standby period, but the condition was not detected. As a minimum, the test and evaluation report should discuss this situation and the results of the various assumptions.

Realistic evaluation of availability requires a realistic view of the logistic delay times.

While operational availability is considered primarily a function of reliability and maintainability, other delay times (e.g., logistic, administrative) involved in restoring the system to an operational condition must be addressed. In the test environment, availability may be optimistically measured if delay times experienced during testing are not representative of delays experienced in operational units.

Standard support factors must be realistic.

Beyond the unit level, the logistics supply system generally is not part of the operational test, and the supply support system is at least partially artificially represented. As a result, it usually is not possible for the test activity to provide meaningful measures for logistics delays that are part of the availability measurement. The Services commonly use "standard factors" to represent delays or resupply times that are included in the estimates for operational availability. To provide a basis for the operational evaluation, these factors should be compared with the delays or supply delivery times that currently are being experienced in actual operation or that are considered reasonable to expect in combat or wartime situations. How realistic these factors are can determine, to a large degree, the validity of the availability estimates that result from OT&E. In many cases, these may be the driving factors for system availability.

Availability and reliability may be traded-off for some systems.

The level of reliability and its relationship to the level of availability may vary significantly from system to system. A system may have a very high availability requirement that is achieved with only moderate reliability and excellent maintainability (e.g., very short repair times). Other systems may have very high mission reliability requirements, but be less sensitive to the level of availability (e.g., numerous items are procured to support a few critical missions). Another system might have high reliability but only moderate availability because of the considerable maintenance time that is required after each failure. It is important to understand the relationship of reliability and availability for each system.

Depot level support should be considered.

In OT&E, the contractor usually performs depot level support for the contractor-furnished equipment (CFE). Organic repair times may be different from those experienced by the contractor, and have a significant effect on worldwide system availability. The suitability assessment should assess the soundness of any planning factor that incorporates the contractor's estimates for depot level repair. The organic repair of government-furnished equipment (GFE) included in the system also may affect availability.

Reliability is a characteristic of the system that relates to how and when failures occur. Reliability is defined as

the duration or probability of failure-free performance under stated conditions.

Mission reliability is defined as

the ability of an item to perform its required functions for the duration of a specified mission profile.

An additional definition is

the probability of success for single-use items, such as rounds of ammunition.

Reliability has a number of different aspects that should be considered. The two most important are termed "mission reliability" and "logistics support frequency."

To describe accurately the mission reliability of an item, the item's required functions, failures of these functions, and the details of the mission profile must be defined. However, mission reliability does not address all the failures that may occur in a system. For example, to achieve high mission reliability requirements, systems are designed with redundancy, i.e., multiple items that are capable of performing identical or similar functions. Failures within these areas of redundancy do not necessarily cause the system to fail to perform mission-essential functions. Failures also may occur in areas that are not related to mission-essential functions. (Weapon systems usually have features that relate to crew comfort or ease of use. Failures in these areas generally do not prevent performance of mission function.)

Logistics support frequency (or logistics reliability, as it is sometimes labeled) considers all calls for the use of logistics resources, whether or not there is an inability to perform the mission. During OT&E, the logistic support frequency must be evaluated to determine if the level of logistics that is demanded by the system is compatible with the planned logistics support.

Since mission reliability considers only critical failures, the mission reliability of a system normally is higher than the logistics reliability. For some items under test, one reliability measure may be more critical than the other, but both must be considered when planning for operational testing. For these reasons, it is important not to confuse the requirements or the measurements of the two areas.

PARAMETERS

Mission reliability is expressed in the classic sense as the probability of performing a mission, without failure, under specified conditions. This parameter is used for systems that perform repetitive missions, such as aircraft or tanks. For systems or items that are used only once, e.g., "single-shot devices," such as missile systems, reliability can be expressed as a ratio of the number of successes to the number of total attempts. These two parameters are expressed as:

$$\text{Mission Reliability} = \begin{array}{l} \text{Probability of completing a mission of} \\ \text{X hours without a critical failure,} \\ \text{under specified mission conditions} \end{array}$$

$$\text{Probability of Success} = \frac{\text{Number of successful attempts}}{\text{Total number of attempts}}$$

Mission reliability also can be stated as Mean Time (Miles, Rounds, etc.) Between Operational Mission Failures (MTBOMF). This parameter can be used for continuously operating systems, such as communications systems, or for vehicles or artillery.

$$\begin{array}{l} \text{Mean Time Between} \\ \text{Operational Mission} \\ \text{Failures} \end{array} = \frac{\begin{array}{c} \text{Total operating time (e.g., driving time,} \\ \text{flying time, or system-on time)} \end{array}}{\text{Total number of operational mission failures}}$$

Another parameter that sometimes is used in place of MTBOMF is Mean Time Between Mission-Critical Failures (MTBMCF), which has a similar equation, substituting the term "mission-critical failures" for "operational mission failures." In both cases, the definition of what constitutes a "mission failure" must be clearly documented for the applicable system.

The parameter "Mean Time Between Failures" (MTBF) is being used less frequently during operational test and evaluation. One reason for this is the confusion that may exist between the use of MTBF as a technical parameter in contract documents or in DT&E, and the use of a similar parameter as an OT&E parameter. (The trend in the Air Force has been to reserve MTBF for DT&E and contract purposes, and to use other parameters with operationally oriented definitions in OT&E.) If MTBF is stated in a TEMP or OT&E plan, the definition of what is considered a failure must be included (or documented in a reference) in sufficient detail to ensure that it includes all operational influences, not just system or component design problems.

Logistics support frequency is measured as the time between events requiring unscheduled maintenance, unscheduled removals, or unscheduled demands for spare parts. These events are considered whether or not mission capability is affected. Logistics support frequency can be expressed as Mean Time Between Unscheduled Maintenance (MTBUM).

$$\begin{array}{l} \text{Mean Time Between} \\ \text{Unscheduled} \\ \text{Maintenance} \end{array} = \frac{\text{Total operating time}}{\begin{array}{c} \text{Total number of incidents requiring} \\ \text{unscheduled maintenance} \end{array}}$$

If MTBF is used as a measure of logistics support frequency, the definition must include any appropriate maintenance events that are not the result of failures such as preventive maintenance actions, inspections, calibrations, or no-fault-found actions. If these nonfailure-caused maintenance actions are not included in the calculation of MTBF, then the MTBF may be significantly higher than a measurement of the MTBUM and is not a measure of logistics support frequency. Other logistics support planning such as manpower, spares, or the amount of test equipment will be understated if MTBF is assumed to be the parameter that determines the need for these resources.

Another logistics support parameter that is indirectly a measure of reliability is Mean Time Between Removal (MTBR). This parameter is used to measure the frequency of failures or maintenance actions that require some item of equipment to be removed from the end item. MTBR indicates the demands on the supply system for replacement items and for repair processing at the intermediate or direct levels of maintenance.

KEY POINTS

Reliability parameters should be defined early in a program.

By Milestone I, the critical reliability parameters should be set forth by the user or user representatives in the early planning documents. These descriptions are the basis for considering reliability and in relating it to other system parameters.

The system's operating modes can drive reliability.

System operating modes must be considered in determining the relative importance of mission and logistics reliability. Defense systems may be continuously operating, perform repetitive missions, or be one-shot items. The reliability parameter must be selected according to how the system is employed. Systems that are continuously operating may be measured by Mean Time Between Operational Mission Failures (MTBOMF). Systems that perform repetitive missions may also use this measure, or the probability of completing a mission without a mission failure.

Firm reliability requirements are essential.

Firm requirements must be established in the Service acquisition documentation for each of the applicable reliability parameters before Milestone II. The requirements have meaning only if they include a detailed, comprehensive failure definition that is consistent with what one would expect from an operational user. If there is an unusual exception (e.g., failures caused by crew error are not included), this exception must be clearly identified in the TEMP and the OT plans.

Reliability measurements can require lengthy test periods.

High reliability systems with lengthy times between failures require lengthy test times in order to gather sufficient test experience to produce meaningful measures of the system's reliability. If extended periods of test time are not achievable, then the OT&E must be structured to use other methods or sources of data to evaluate reliability. Modeling and simulation might be used to focus the operational testing on subsystems of greatest risk or criticality. The data of technical testing might be used to supplement data that are obtained during OT. In some instances, the only solution is to be satisfied with reduced confidence levels, since the cost of the test systems (e.g., complex missile systems) is too high to permit additional test articles to be expended.

Assumptions are made in reliability test planning.

Long test periods are needed to accurately measure reliability of high reliability systems. Sometimes a greater number of items are tested for a short time, instead of a few items for a longer time. For example, a good test program for a system with a MTBOMF of 1000 hours might be to test some number of systems (e.g., four) for 1000 hours each. Because the long test period needed to obtain 1000 operating hours is not possible, the test is planned for twenty systems with 200 hours each. Although this will result in the same number of total operating hours, there are inherent risks. The DOT&E evaluator will need to determine the likelihood of significant "wear-out" failure modes -- are there significant failure modes that will not be seen until after the 200 proposed hours of testing? Assessing this risk requires an examination of the system. The responsible Service should demonstrate that the risk is acceptable by presenting other test data that demonstrate that significant "wear-out" failure modes have not occurred in longer duration testing, and are unlikely to occur in operation. If this cannot be demonstrated, then alternative test approaches should be considered, e.g., testing a few items for a longer period, having test units tested for a longer period in other test phases, etc.

Early OT may give the first realistic view of system reliability.

Initial operational testing can provide insight into the reliability potential of the system. While it is difficult and time consuming to verify that a high level of reliability exists, the discovery of reliability deficiencies is not difficult if the system is truly deficient. An operational assessment may identify areas that potentially could become major deficiencies. A critical error at this stage is committed when the reliability data that are the result of early testing are not accepted -- failures are "explained away" and managers' projections that accept high reliability turn out to be too optimistic. On the other hand, the reliability and failure experience early in a development program may not be representative of the production configuration; the results indicated by early Developmental Test (DT) data may change significantly by the time the system reaches maturity.

If there are any deviations from the planned production configuration, they should be understood.

To evaluate the reliability of a system in its intended operational environment requires that the items being tested have a reasonable degree of agreement with the planned production configuration. All significant deviations from the production configuration should be identified.

Reliability measures can have statistical confidence calculations.

One measure of the sufficiency of the test data is the use of statistical confidence calculations. During the test planning, confidence calculations can be used to determine how much testing is needed to yield a certain level of confidence in the results. After the test data have been collected, confidence calculations can be used to indicate how adequate the data were in determining the values for the reliability parameters. A good reference for a discussion of confidence levels and intervals, and how they apply to the test and evaluation environment, is DoD 3235.1-H, "Test and Evaluation of System Reliability, Availability, and Maintainability - A Primer." It provides a reference for the mathematical aspects of RAM and discusses confidence as it applies to planning test programs and evaluating test data. The discussion is based on technical R&M characteristics and may require tailoring when applied to OT&E.

Software reliability is always an issue.

As defense systems become more and more software-intensive, understanding software's contribution to system reliability increases in importance. Software faults and errors can be a major problem during the initial phase of a system's operation, and they should be viewed within the context of what effect the faults will have on the system's performance. Serious faults can result in mission failures and should be treated accordingly. Easily corrected, minor interrupts that cause no system-mission impact may be significant only if the quantity is high enough for the problem to effect operator performance and attention.

Reliability growth programs are used in some DoD programs.

If the program manager has defined a reliability growth program, the projections from such growth programs should not be used as part of the operational evaluation. The potential growth of system reliability, during or after the completion of the operational testing, may not be easy to estimate; it is dependent on resources, dedication, and many technical details. If a projection is needed, it is preferable to use the reliability growth experience from similar systems as a basis for what has been done and what might be done on the system in question. A projection should never be reported as a test result. The test result should be an observed value.

Maintainability relates to the ease and efficiency of performing maintenance. It is defined as

> *the ability of an item to be retained in or restored to specified condition when maintenance is performed by personnel having specified skill levels, using prescribed procedures and resources, at each prescribed level of maintenance and repair.*

There are three important dimensions to the examination of system maintainability. The first is the average corrective maintenance time required to restore the system to its mission-capable condition. This maintainability characteristic gives a view of how long the system will be under repair after mission-critical failures. The average repair time for a system might be two hours. In this situation, the system would be unavailable due to maintenance for an average of two hours after each mission-critical failure.

The second dimension addresses the maintenance time required to restore the system after any failure that requires corrective maintenance. The average time to restore, considering all corrective maintenance, may be longer or shorter than the time for mission-critical failures.

The third dimension to consider is the manpower required to perform the repair function. If it takes two hours for the average repair, there is a considerable difference in the required support resources if the repair requires one technician or three technicians for this two-hour period.

PARAMETERS

A maintainability parameter that addresses the length of time required to restore the system to a mission-capable state is Mean Operational Mission Failure Repair Time (MOMFRT).

$$\text{Mean Operational Mission Failure Repair Time} = \frac{\substack{\text{Total number of clock hours of corrective, on-system,}\\ \text{active repair time used to restore failed systems to}\\ \text{mission-capable status after an operational mission failure}}}{\text{Total number of operational mission failures}}$$

A parameter that addresses the time required to restore all failures is mean corrective maintenance time (MCMT).

$$\text{Mean Corrective Maintenance Time} = \frac{\substack{\text{Total number of clock hours of corrective, on-system,}\\ \text{active repair time due to all corrective maintenance}}}{\text{Total number of incidents requiring corrective maintenance}}$$

Another parameter that is used to address all corrective maintenance time is mean time to repair (MTTR). The OTA MOA (see Chapter 3) has a number of definitions for MTTR. One definition is

$$\text{Mean Time To Repair} = \frac{\text{Sum of corrective maintenance times}}{\text{Total number of corrective maintenance actions}}$$

In addition to the average (or mean) time that is required, there are some systems where it is important to have a view of how many lengthy maintenance actions there will be, and the duration of those actions. In this situation, the parameter Maximum Time to Repair (MaxTTR) may be used. Maximum Time to Repair will give a time that a specified percent of the maintenance actions would be completed. For example, a MaxTTR might be specified for 90 percent of the failures. If the MaxTTR was four hours, this would mean that 90 percent of the maintenance actions would be completed before four hours elapsed.

To use parameters such as Mean Corrective Maintenance Time (MCMT), it is important to define the meaning of "corrective" versus "scheduled" or "preventive" maintenance, and also to define when each of the important time measures start and stop. For space systems and software-intensive systems, the parameter Mean Time To Restore Function (MTTRF) may be used.

$$\text{Mean Time To Restore Function} = \frac{\text{Total maintenance time to restore mission functions interrupted by critical failures}}{\text{Total number of critical failures}}$$

This parameter addresses both scheduled and unscheduled maintenance time.

Maintainability parameters that address the manpower resources required to perform maintenance include Maintenance Manhours per Operating Hour (flight hour, mile, round, etc.). For some systems, the maintainability (and the manpower required) is measured using a parameter of Maintenance Ratio (MR), which is a measure of the maintenance manpower required to maintain a system in an operational environment. It is expressed as the cumulative number of direct maintenance man-hours during a given period of time, divided by the cumulative number of system life units (such as hours, rounds, miles) during the same period. There could be a measure of MR for each maintenance level, and/or a summary for a number or all levels. The levels of maintenance are defined and labeled differently in each of the Services (see page 5). The man-hours considered usually include all types of maintenance, e.g., corrective, scheduled, preventive.

Maintainability also may include an examination of any automated system diagnostics. The label "integrated diagnostics" addresses all forms of diagnostics systems, including automatic, semi-automatic, and manual, built-in or stand-alone, in an integrated fashion. All faults and failures must be diagnosed by some method. The integrated diagnostics systems should be designed and tested to meet this need. Integrated diagnostics is discussed in section 3.2.

The maintainability of the software of certain systems also can be a critical issue. This topic is discussed in section 3.5.

KEY POINTS

Maintainability measurement requires a reasonable number of maintenance events.

Limiting the test period can limit the applicability of the maintainability results. To obtain a meaningful evaluation of system maintainability requires that the test data include a relatively good cross-section of maintenance actions that are expected to occur during the system's operational service. Examining only a few failures may exclude insight into a large portion of the maintenance training, maintenance documentation, etc. Sampling a number of maintenance actions can result in a maintainability evaluation that is significantly different than the actual capability of the system. One solution is a properly structured maintainability demonstration, but such demonstrations must be properly planned to produce results that are valid for supplementing operational test results.

Maintainability demonstrations can be used in OT&E if they are realistic.

A maintainability demonstration is an activity wherein maintenance tasks are caused to be performed and the personnel, test equipment, tools, etc., involved in the maintenance are evaluated by observation. Proper operational maintainability demonstrations must be structured so as to present a representative operational environment. Maintainability demonstrations routinely are used as part of the contract verification requirements imposed on the developing contractors, but the demonstration methods used there are not adequate for OT&E because they lack operational realism.

If maintainability demonstrations are used as part of the OT&E, the plan for these demonstrations should address the cross-section of maintenance tasks that are used, the personnel used in performing the maintenance, and the conditions surrounding the tasks used in the demonstrations. Maintenance events or tasks that are performed in operating units usually include events that are not the result of failures. Further, there are poorly documented symptoms, false alarms from the diagnostics system, and other types of tasks that may not be included in a task list that is provided by the contractor or the government program manager.

The best source of a realistic task list would be the product of having senior maintenance technicians from the relevant operating units modify the contractor's task list, based on maintenance tasks required for similar systems. These task lists can be used to identify pre-faulted modules that have been used as part of maintainability demonstrations. The personnel used in the demonstrations should be as representative of the expected operational personnel as possible.

The conditions attending the demonstration also should be realistic. It should not be conducted in a laboratory if the actual maintenance will be done in the field, in all weather, and under 24-hour-a-day lighting conditions. If maintenance tasks are to be performed while wearing chemical, biological, radiation (CBR) protective clothing, then the use of this clothing should be considered when planning the demonstration. The proximity of support assets, such as support equipment, documentation, etc., also should be representative of what might exist in an operating unit.

Built-in test equipment and other diagnostics systems must be properly tested and false alarm rates documented.

Built-in test equipment and other diagnostics systems must be tested properly, as they too may be fundamental to maintaining the system. This testing must not be glossed over, as is often the case. False alarm rates frequently are not discussed in test documentation and therefore make it very difficult to evaluate the BIT responses that occur during OT (see section 3.2).

Any routine scheduled or preventive maintenance should be carefully examined during OT.

The total requirement for scheduled or preventive maintenance can be significant for some systems. This significance may exist either in the total system downtime to perform the maintenance, or in the total number of manhours required. Routine maintenance events, such as changing oil in generator power supplies, should be examined for adequate accessibility. The "reasonableness" of the estimates for the total downtime or manhours expended in these areas should be examined. The time required for each of the significant scheduled maintenance tasks should be evaluated as part of the OT.

The time for off-equipment repairs can be significant.

One factor often overlooked in requirements and test documents is the criteria for off-system (or off-equipment) repairs. Mean time to repair (MTTR) usually applies to the time required to return the system to a mission-capable state. This may involve a change in a major system component. The time to repair applies to the time needed to isolate the failure to a component and to change the failed component, but does not involve the actual repair time to return the component to a ready-for-issue status. MTTR can be manipulated by a "shotgun maintenance" approach (e.g., removing and replacing the three or four most likely failure candidates), but such actions may go undetected if the removed parts are not tracked through the next higher level of maintenance. Further, this approach causes serious logistics concerns due to the quantities of good units being processed in the repair pipeline, tying-up maintenance resources later labeled as "Retest Okay" items. It also may have a major affect on the number of spares and how they are positioned.

An evaluation that addresses only on-equipment MTTR fails to consider the impact on the maintenance organization from the manpower and support equipment required to actually repair the item, and it does not consider the impact on the logistics system of the material needed to make the eventual item repair. Off-system repairs should be evaluated to determine the potential for unexpected levels of support costs. The inability to logistically support the system once fielded can be significant. To avoid this, off-equipment repair should be addressed in the requirements and OT&E documents.

Unique maintainability characteristics or requirements should be identified and included in the OT.

The system being tested should be examined to determine if there are any unique maintainability characteristics or requirements. Examples might be maintenance of nuclear hardness features or special features for battle damage repair. These unusual features should be considered when planning the maintainability activities for the OT.

19

Interoperability is defined as

the ability of the systems, units, or forces to provide services to and accept services from other systems, units, or forces, and to use the services so exchanged to enable them to operate effectively together.

In the context of operational suitability and OT&E, interoperability addresses the ability of the system to be used in concert with other types of systems and other systems of the same type that are necessary to accomplish the required mission or missions.

Interoperability is frequently considered to have an effect on both the suitability and the effectiveness of the system. Operational testing must be focused on those supporting or companion systems that are essential for the system under test to meet its operational requirements. During the early test planning, the critical companion systems must be identified and documented by the Service in the TEMP. The quality of the interoperation that must be examined by operational testing must be defined in detail in the OT&E test plan prepared by the OTA.

Interoperability recognizes that, for the system to perform its mission, there are functions that are performed by two or more items in concert. For example, data and communications systems must have the technical capability to interface and exchange data. Issues for these systems are usually resolved during technical testing, but operational testing may result in additional confidence in the interoperability, or may provide additional and more realistic situations under which to assess the operation of the systems.

PARAMETERS

Interoperability usually is evaluated in a qualitative manner. However, there may be aspects that are described quantitatively and therefore contribute to the presentation of the system's interoperability status.

One way to describe the interoperability of a system being tested is to discuss what limitations the system imposes on operations when it is used with other systems. This would entail preparing the following:

- A list of those systems that will require special procedures when operated simultaneously with the system under test, and

- A list of those systems whose modes of operation must be changed when in the presence of the tested system.

KEY POINTS

Supporting or companion systems need to be identified in the early versions of the TEMP.

The identificatizxcv nm0on of supporting or companion systems is one of the keys to conducting a successful evaluation of the system's interoperability. By Milestone I, the systems and equipments that will have critical interoperation with the system under test should be identified and documented, by the Service, in the appropriate acquisition documents. The early versions of the TEMP should highlight these companion systems and the need to have the required test assets to verify interoperability.

The consideration of companion or supporting systems also should address other systems that are under development.

This consideration should include not only existing systems, but also companion systems that are being developed at the same time as the system being examined. There are frequently problems in acquiring test assets when systems are in the development stage. In such a case, conducting dual tests of two systems that are both being developed may be difficult, but the OT&E planning must address the need for examining any critical interoperability. If such dual testing is not possible, then the evaluation must discuss the likelihood of potential problems and the limitations to the OT&E because the two systems were not tested simultaneously.

Maturity of supporting or companion systems must be understood.

The relative maturity of the supporting or companions systems must be part of the assessment of what is needed during the OT. If these other systems are not mature enough to provide a realistic level of the planned support, the OT may provide invalid answers. Also, the likely maturity at fielding should be assessed. If the suitability of the system when fielded is dependent on supporting systems, e.g., targeting information systems, and the supporting systems are unlikely to be available to provide the required support, then this potential deficiency needs to be highlighted to the decisionmakers.

Determination of adequate suitability depends on the performance of the supporting systems.

When the suitability of the system under test is being determined, the acceptability of the supporting systems also should be part of the judgment. An example might be a targeting system for a missile system: if the targeting system cannot meet its projected requirements, then the missile system can hardly be expressed as being suitable.

Interoperability problems may cause system limitations.

One category of interoperability problems is a situation where a system must be limited in its operation due to the proximity to another system. Examples include a radio transmitter that must be turned off when near another type of radio or communications device; an aircraft that must fly at reduced speeds when in the company of another aircraft type; and a jammer that must be turned off if a radar is to work or if a certain missile is to be fired. Another example is the limitation on the use of viewing devices, low-light-level television, etc., if flares are to be used. Limitations to system operation that are the result of interoperability should be identified during OT.

Interoperability should be addressed in the OT&E prior to Milestone III.

By Milestone II, the TEMP should indicate the manner in which major interoperability areas will be examined under operational testing. The test resources description must indicate what companion or supporting systems are critical to the system in question. The test resource planners need to assure that these companion or supporting systems will be available for the test period.

2.5 COMPATIBILITY

Compatibility is defined as

the capability of two or more items or components of equipment or material to exist or function in the same system or environment without mutual interference.

Compatibility addresses and includes many different areas. It concerns the capability of the equipment in the system to operate with each of the required supporting equipments, e.g., electrical power generation, air conditioning, hydraulic power subsystem, as well as with other elements of the system. It also addresses the interface with logistics support items, including test equipment, servicing equipment, maintenance stands, handling equipment, and elements of the transportation systems (see section 2.7). Compatibility includes physical, functional, electrical and electronic, and environment conditioning areas. Human factors (covered in section 2.13), environmental factors (section 2.3), and electromagnetic environmental effects (E3) (section 3.4) also are compatibility considerations.

Physical compatibility involves attachment pins, connectors, the interconnecting wires, cables, alignment, and mechanical linkages. Physical compatibility may involve the ability to install the item in its assigned location, physical clearances, and item volume. These physical characteristics involve compatibility with other elements of the operational system, as well as equipment that is part of the logistics or maintenance environment. Electrical or electronic compatibility considerations include voltage, current, and the frequency for systems using alternating current. For radio frequency or visible light interfaces, a basic consideration is the frequency of the transmitted signal. Other factors considered include bandwidth, frequency hopping patterns, and signal polarization. Environmental conditioning considerations address the compatibility of heating and cooling subsystems. The cooling to be provided for electronics items must be consistent with the requirements of the system.

PARAMETERS

Compatibility parameters involve the measurement of many different aspects of the system's characteristics, as well as compatibility by function. While much of the detailed compatibility testing is the domain of DT, there is a need to monitor and assess compatibility during OT. Additional or expanded environments usually are present during the conduct of a realistic OT, and therefore there is some likelihood that problems not seen during DT may be uncovered in the operational testing.

Some of the parameters that may be considered during DT and/or OT include:

- Physical - Attachment pins and connectors, alignment, physical dimensions, volume, and weight.

- Electrical - Voltage, cycles, power profile or stability, and surge limits.

- Electronic - Frequencies, modes, rates, control logic, and telemetry.

- Environmental Conditioning - Heating, cooling, shock and vibration protection.

- Software - Formats, protocols, and messages.

- Hardware/Software - Conventions, standards, timing, sequencing, sensing, and control logic.

- Data - Rates, inputs, characters, and codes.

KEY POINTS

DT results may help focus OT planning.

Compatibility requirements should be monitored during the early development stages of the acquisition process to ensure all compatibility areas and issues are addressed. DT results should be tracked to assist in the OT planning and to avoid duplication of testing efforts.

Early operational testing may indicate unforeseen compatibility problems.

The early OT phase should include a compatibility issue to ensure that operational considerations introduced into the testing at this point are not causing problems that were not observed in DT. Identification of potential problems not anticipated by the designer could include electrical power variations, unexpected electrical interference, or the need for additional air conditioning.

Nominal operations may not expose incompatibilities.

Nominal operations may not expose interference or incompatibility problems, and special tests may be required to test the system in various modes and operational extremes to detect potential interference.

Operational test personnel must address the needs for any special resources/systems that are required for compatibility testing.

If special facilities, instrumentation, and simulators for compatibility are required for OT, advanced planning is required. These requirements must be discussed in early versions of the TEMP. Failure to do advanced planning may result in a delay to the operational testing, or in conducting the test without completing some of the objectives.

Modifications or upgrades may introduce compatibility problems.

The addition of new or advanced capabilities to a weapon system may introduce the potential for compatibility problems. For example, if a system is upgraded with more advanced computer and electronics systems, the original environmental equipment may be unable to provide adequate cooling to the new system. As a result, the new electronics will operate in higher temperatures and be less reliable than projected when used by the operating units. A similar situation may result with a ground combat vehicle that is upgraded with heavier weapons and, as a result, has a weight that may be incompatible with elements of the drive train, brakes, etc. The integration of non-developmental items (NDI) into a system also may introduce compatibility problems.

Compatibility of procedures can be a factor in system performance.

The compatibility of two systems may depend to a large degree on the procedures being used and how the procedures are followed. During the testing of one system, it was discovered that the main system was not compatible with the command and control system. The system was fired and controlled in a fully automated manner, while the fire support system was manually operated. The crews that were to make the manual-to-automated translations could not communicate inside some of the shelters because of the noise level that resulted from the support equipment (400-cycle cooling fans, and turbine generators).

2.6 LOGISTICS SUPPORTABILITY

Supportability is defined as

the degree to which system design characteristics and planned logistics resources, including manpower, meet system peacetime readiness and wartime utilization requirements.

This element addresses the balance between the system's support needs, which are a result of the system design, and the planned logistics support for the system.

Other suitability elements also address aspects of this balance, e.g., reliability, maintainability, manpower, documentation, training, and the like. The scope of logistics supportability within the OT&E documentation is limited to those aspects that are not covered under other topics. It also includes the integrated aspects of the logistics planning. Key items that should be addressed under this element include supply support (the planned numbers and placement of spares and repair parts), test or support equipment for all levels of maintenance, and planned support facilities.

Some systems go into operational testing without a plan for specific numbers of spares or repair parts. This situation becomes a test limitation. The problem is due in large part to cost and manufacturing constraints imposed either on or by the developer. In those cases, the contractor provides a portion of the required system support package.

PARAMETERS

Logistics supportability is most often evaluated in a qualitative manner, although there may be quantitative factors that are used in some of the elements of supportability. For example, supply support might be assessed by examining parameters such as percent of items in local supply assets, fill rates, etc. These parameters are then judged qualitatively to arrive at the assessment of the entire subject of logistics supportability. The evaluation of supportability is intended to consider all these subordinate factors, or considerations, and provide a composite evaluation of the balance between the support that is needed and the support that is planned.

In some test programs, the area of logistics supportability also is used to cover other suitability areas (e.g., transportation, manpower supportability, documentation, or training) that are not significant enough for a particular system to warrant an individual test objective.

KEY POINTS

Early ILS planning can be assessed as part of logistics supportability evaluation.

Early activities in Integrated Logistics Support (ILS) planning may begin before Milestone I. Some aspects of Logistics Support Analysis (LSA) also are conducted in this time frame. Review of the products of these activities may be key to any early operational assessment of operational suitability. These early activities also will give some information on the criticality of the support aspects of the system. If the system is to have a unique support concept that will succeed only under certain system reliability and maintainability levels, then this might be identified as an operational suitability Critical Operational Issue (COI). Such an examination will begin to focus on the critical aspects of the suitability that need to be addressed in the OT&E

planning. One example of potential support problems is a system that is intended to support a number of different weapon systems -- within the context of one weapon system, the planning may be accurate, but for others or for the total support requirement, the planning may not be correct.

ILS planning can provide the basis to assess the planned logistics support.

An early assessment of logistics supportability can be made by reviewing the status of the logistics support planning. One portion of this activity should be a review of the Integrated Logistics Support Plan (ILSP), or of the results of the logistics support analysis. How complete and well the planning is done determines, to a large degree, how acceptable the support of the system will be. Modeling and simulation (see section 3.1) also may be used to analyze the ability of the system to meet some of its suitability requirements with the planned level of support.

Test planning must address the support for the items under test.

As the operational test is planned, the support assets (e.g., spares, test equipment, support equipment) must be identified in the TEMP and in the detailed test plan and must be in place for the conduct of the operational test. If these support assets are not available, it will be more difficult to assess the operational suitability of the system, and it will be impossible to evaluate the performance of the support items.

Operational test data should be compared to the ILS planning factors.

Assessment of the ILS planning should continue as part of the OTA's activity to support the OT&E report for the Milestone III decision. Operational test data should be compared with the ILS planning factors and evaluated to determine if the planning factors reflect the real needs of the system. If the test data contain demand rates (e.g., MTBR) from the operational test period, these rates should be compared with the rates that were projected in the logistics planning. Incompatibility of the planning and the system parameters is one of the major causes of systems being unable to meet their availability requirements during the first stages of operational fielding.

Supportability of software should be considered.

For those systems that have significant software elements, the support of the system's software can be an important factor in the ability to provide logistics support for the entire system. Planning for support of software includes both manpower and equipment resources. The personnel who will be used to maintain or upgrade the software during its operational use must have adequate documentation to allow them to perform their function. Evaluations during OT&E can provide insight into the adequacy of the planning for the support of the software. (Software supportability is discussed in more detail in section 3.5.)

Supply support during operational testing may be unrealistic.

Some test programs provide an "iron mountain" of spares and repair parts to optimize the use of valuable test range times, test forces, instrumentation, etc. In these situations, the evaluation of test results must compensate for the unrealistic conditions in the logistics support of the systems under test.

2.7 TRANSPORTABILITY

Transportability is defined as

the capability of materiel to be moved by towing, self-propulsion, or carrier through any means, such as railways, highways, waterways, pipelines, oceans, and airways. (Full consideration of available and projected transportation assets, mobility plans and schedules, and the impact of system equipment and support items on the strategic mobility of operating military forces is required to achieve this capability.)

System requirements and employment methods may dictate that specific transportation modes be used for deployment purposes for some deployment scenarios. Assessment of these different modes of transportation should be accomplished by the developing agency. Other areas of attention include the need for any unusual transportation or handling equipment. The compatibility with transport aircraft, ships, or any vehicles that are essential for the system to arrive at its destination and then perform its mission also must be addressed. This compatibility includes physical dimensions and clearances, tie downs, and load capacity. Routine transportation and mobility movement of spares and support equipment also should be considered. The capability of the item to be included in any required at-sea replenishment or amphibious operations should be addressed. These elements include the ability of the system to be transported by the planned means or within the intended transportation capability of the DoD.

Transportability also may include the "deployability" of a system or equipment that is deployed with combat units to the combat area, and the "portability" of items that are carried by user personnel during use. Compatibility with the using personnel also must be assessed. Personnel who will prepare and move the item as part of its transportation must be physically capable of performing the required tasks.

PARAMETERS

Parameters associated with transportability must address the characteristics that will allow existing transportation assets to move and transport the equipment as needed to support the operational mission. If new transportation assets are planned specifically for this system, then the evaluation should address the acceptability of these new assets. Some parameters associated with transportability are:

- Does the using organization have the provisions for handling and transporting the system?

- Can the system be transported to the theater by the preferred means?

- Can the system be moved adequately within the theater of operations?

- Are the dimensions and weight within the required limits for each possible transportation mode that will be required to move the equipment?

KEY POINTS

Unique transportability requirements should be identified.

Initial system transportability requirements should be specified in the early system planning documents, and the acquisition and using agencies should assess these requirements against the capabilities of existing transportation assets. The role of OT planners is to examine these requirements and determine if test resource assets (e.g., cargo aircraft, rail cars) are needed for OT. The need for special transportability test events also should be addressed. If there is a critical transportability issue, this needs to be identified in an early TEMP.

Transportability of the system should be verified as part of operational testing.

If the system is a major rail- or air-transportable item, for example, a tank or a helicopter, the compatibility with the transporting means should be verified. Often, the DT will examine this compatibility, but from a technical standpoint, i.e., does it "fit?" While an argument can be made that the transportability requirement has been verified by this DT evaluation, there are operational factors that need consideration. Just because contractor engineers are able to load the system into the transporting aircraft on a clear day in good weather does not mean that the using troops can perform the same task under all weather and lighting conditions. In the case of manpack items, the ability of the person to carry the item may be proven, but can the person who is carrying the item still perform the assigned mission, or is there some negative impact on combat effectiveness. The OT examination is directed more at "can it be done by the normal user troops under the conditions predicted for the using organization." Operationally realistic conditions can yield results different from those produced by the DT.

All projected areas of operations should be included in the transportability assessment.

Due to weight, dimensions, and system characteristics, the transportability of the system may be limited in certain geographic areas of operation. For example, the dimensions of a tank may be compatible with U.S. rail transportation, but the use of rail transportation in other countries will be limited because rail widths and capability differ from those in the United States. Systems with extensive global commitments must be analyzed very carefully to ensure that all transportability requirements are understood and can be met. If the system has a unique transportability requirement, it should be made part of the system planning documentation and considered for examination as part of the OT.

Transportability should include the movement of the system into combat locations.

The system must be moved into and within a theater of operations consistent with its mission. This issue may deal with airplane, train, or ship loading and internal or external helicopter loads. The examination should address the ability of the transporting system to carry the load, and any impacts on maneuverability once loaded. It should ensure that the weight and dimensions of the new system can be supported by the transportation network and current bridging (to include tactical bridging) in the required operational environment.

Testing of systems after being transported can be critical for some systems.

For some systems, operational testing should be planned to verify the fact that transporting the system has not degraded its capability. Realistic scenarios of preparation for transport, the actual transport of the system, and set-up for operation all should be included in the operational testing scenario. The importance of this activity varies from system to system, and should be included in the testing requirements if the requirement is judged to be significant.

2.8 DOCUMENTATION

Documentation is a portion of technical data and information that is part of every system.

For the purposes of OT&E, documentation comprises operator and maintenance instructions, repair parts lists, and support manuals, as well as manuals related to computer programs and system software.

The ability to operate and maintain new and advanced systems can be highly dependent on the completeness and accuracy of the documentation that is provided with the system. For complex systems, the operator and maintainer documentation can make the difference between success and failure of the system.

During the documentation development process, usually during the full-scale development phase, representatives of the user organizations generally will conduct a validation of the documentation. The validation process addresses the ability to locate procedures and tasks, as well as the need for any additional tasks required to support maintenance operations. Documentation clarity, accuracy, and ability to support projected skill levels also is validated. Cautions, warnings, and advisories are reviewed to ensure that they are appropriate for incorporation in the manual, and are checked to ensure that they are accurate, clear, and easily identifiable to the reader. Preventive maintenance checks, services, and procedures also are validated.

The validation process generally is accomplished in two phases. Developmental testing personnel usually perform the first phase, and determine if the drawings, figures, specifications, and procedures are technically correct. The second phase usually is done by operational testing personnel. They determine if the maintenance technician and operator can understand and correctly perform the procedures outlined in the documentation. This examination may be performed in conjunction with other suitability evaluation activities, including data collection for maintainability, training, and human factors evaluations.

In addition to any "formal" validation tasks, the documentation should be also part of the activity on other OT tasks. For example, when operations or maintenance is performed during the OT, the documentation that is used should be assessed for its completeness, accuracy, ease of use, etc. The OT&E plan should discuss how the results of these naturally occurring documentation assessments will be recorded.

PARAMETERS

Documentation evaluation is primarily qualitative in nature. There are some quantitative parameters that might contribute to organizing and managing the assessment of the documentation. Three examples follow.

Percent of Critical Tasks or Procedures Available: Clearly, the weakest documentation procedure is the one that does not exist. This parameter indicates how complete the documentation is at the time of OT. Documentation can be made available in various forms; e.g., draft, "blue line," final deliverable. The percentage of each form that is provided to the operational test activity can be a good indication of the status of the documentation development.

Percent of Critical Tasks or Procedures Validated: For the total number of tasks or procedures in the operating or maintenance manuals, this represents the percentage that have been validated. It can be applied to maintenance, operator, or other support-critical tasks or procedures, and it should approach 100 percent as the system nears the production decision.

Percent of Erroneous Procedures or Tasks: For the total number of tasks and procedures demonstrated, this is the percentage that are considered to be erroneous. Other similar parameters can be developed to highlight the number of tasks that are unclear and tasks that have too much detail, insufficient detail, etc. The parameters will depend on the manner of collecting this information at the operational test site.

KEY POINTS

Documentation should be available for the operational test phase.

The documentation development and assessment schedule must be compatible with scheduled operational test time frames. The early assessment of the documentation preparation program is one form of early suitability assessment. A good process, with adequate schedule and resources, may produce good documentation. A poorly scheduled program, or one with inadequate resources, probably will yield poor results. At a minimum, preliminary documents should be available for the operational testing phase, even if the final documents are not ready.

Documentation may not be available for the operational testing schedule.

While the best test of documents is to use them during OT, the delivery dates for the documentation may make it difficult to evaluate the final product during OT. The effect of any schedule shortcomings should be known well in advance and arrangements made for work-arounds, use of draft documents, or other alternative evaluations. For accelerated programs, the OTA should identify, and include in the OT&E plan, alternative methods for achieving the evaluation of this area. Delays in availability of essential technical manuals can cause a test to experience disruptive delays or, more significantly, result in an improper evaluation of the planned support system or system reliability, availability, and maintainability.

Assessment of documentation may be in a separate test phase.

The assessment of the documentation usually requires a separate documentation test phase to determine its adequacy, prior to using the documentation as part of system operational testing. This testing should stress the use of military personnel skills, tools, facilities, and support equipment that are planned for use in the support environment when the system is fielded.

Only a sample of the operation, maintenance, and support tasks in the documentation may be naturally occurring in OT.

One of the difficulties in documentation assessment is that it needs to address a relatively large percentage of the operating and maintenance tasks at the organizational and intermediate (direct support) levels, as well as a sampling of maintenance tasks at the general support level. It may be difficult to evaluate voluminous documentation when the test period is relatively short. The use of system documentation during operational testing will provide data on its acceptability in a narrow range of circumstances. One alternative is to use data from other sources, e.g., maintenance during DT and maintainability demonstrations (see section 2.3). Any critical task or procedure not observed adequately during the operational testing should be validated in a separate scheduled event.

2.9 MANPOWER SUPPORTABILITY

Manpower supportability is defined as

the identification and acquisition of military and civilian personnel with the skills and grades required to operate and support a materiel system over its lifetime at peacetime and wartime rates.

Within the context of operational test and evaluation, manpower supportability takes into consideration the numbers, skill types, and skill levels of the personnel required to operate, maintain, and support the systems in both peacetime and wartime environments. Manpower supportability, therefore, is closely related to training requirements, human factors, and maintainability.

Increases, deletions, and changes to the force structure may be required to place the system in its operating units. Documents are developed which indicate the projected manpower requirements and skill codes and grades that are necessary. The OT objective here is to assess if the projected levels are adequate to operate and support the system.

The determination of the number of manpower spaces required is based on the various scenarios in which the system will be used. The number of spaces depends on the system having the projected reliability and maintainability, and the support resources, diagnostics, test equipment, etc., as planned. Shortfalls in these areas can have significant impact on the required manpower. If the projected manpower is inadequate to support the system, then significant problems could occur when the system is fielded.

PARAMETERS

The general parameter for manpower supportability is the number of personnel required to man the system when it is employed. It addresses operating, maintenance, and other support personnel, and their required skills and training. Other parameters that might be used include the following three:

Crew Size: The number of people required to operate the system and perform the tasks required in each speciality and at each skill level, and to use the system in the intended scenario(s).

Maintenance Ratio: The ratio of maintenance manhours per operating hour or life unit (see section 2.3). This measure allows the comparison of the projected maintenance workload and the workload demonstrated during operational testing. On- and off-system Maintenance Ratios (MRs) are estimated for each level of maintenance and for each skill code. MR criteria typically are found in the requirements document.

Current Crew Size and Maintenance Ratio: A comparison of the system's manpower requirements with those of the current system can provide a meaningful measure of the criticality of the manpower supportability.

KEY POINTS

Manpower supportability includes examination of the operating crew.

While manpower supportability may highlight the support aspects of manpower (e.g., maintenance crews and skills), the operating crew is an important part of the evaluation. The OT&E test plan always should indicate that the operating crew size, skills, etc., will be part of the evaluation.

Manpower deficiencies actually may reside in other suitability areas.

On some systems, it may appear that the manpower estimated is inadequate to operate or maintain the system. If the system is experiencing shortfalls in expected reliability, maintainability, or diagnostics, or has human factors problems, then the deficiency may be more correctly assigned to these areas.

Manpower planning for OT&E should not include "Golden Crews."

The term "golden crews" is used to identify operating/maintenance personnel who have skill levels higher than those of the field crews. The personnel who man and maintain the systems during the operational testing should not be of such a high skill level that the test results are invalid. The thrust of OT&E is to see if the system is suitable in the hands of "troops" who are representative of the intended operational users. Realism also may be lost if the personnel in the testing organization have received a greater number of exposures to the tasks than will be the case with the planned user personnel. An example would be the use of hand-held ground-to-air missiles -- if the intended users will never have the opportunity to fire a live round during their training, but only have exposure to simulators, then it is unrealistic to use personnel in the OT who have had experience with a number of firings during DT, or other testing of this system, or similar systems.

Skill levels and numbers may be hard to evaluate.

The complement of personnel that is used during some operational tests may have somewhat higher skill levels than are planned for the operational units. These higher skill levels are justified by the OTA as necessary because it takes higher skill level personnel to recognize deficiencies in the system. In this situation, the evaluation of the manpower resources needed to operate and maintain the system must consider the skills that were present during testing. The OT&E plan must include an adequate evaluation procedure.

Proper manning levels for systems are critical for efficient operations.

The OTA must carefully evaluate the manning levels of units who will field new systems. Improperly manned equipment will result in poorly operated and maintained systems. During the fielding of one system, it was discovered that the system was not properly manned; additional personnel were required to allow units in the field to be more self sufficient. On another major system, the proposed maintenance organizations did not have personnel or a supporting organization to maintain the required radio frequency signal management system. This lack of manpower would have resulted in those maintenance tasks being transferred to a higher maintenance level, thereby reducing the efficiency of the unit.

Training and Training Support are defined as

the processes, procedures, techniques, training devices, and equipment used to train civilian and active duty and reserve military personnel to operate and support a materiel system. This includes individual and crew training; new equipment training; initial, formal, and on-the-job training; and logistic support planning for training equipment and training device acquisitions and installations.

During the OT&E, the planned training program, along with any training devices and equipment, should be evaluated. The training program and supporting materials are developed during various phases of the weapon system development process. The supporting materials include programs of instruction, on-the-job training documentation, training materials, and, when required, training aids and simulators. Training materials are provided for both individual operator and maintainer training, as well as for "collective" training for crews or units.

Operators are trained to perform all critical tasks required to operate the system. Maintenance personnel are trained to perform all critical tasks required to maintain the system. Collective training is provided to system crews whose members are required to perform as a team. All tasks must be accomplished to prescribed training standards. If tasks to be performed are linked to or dependent on other tasks (e.g., firing sequence, or an initialization sequence), all tasks must be performed to standard in a single performance test.

Evaluation of training should address the effectiveness of the training program in providing personnel who can operate and maintain the system. Maintenance training must be analyzed from the organizational through the intermediate level and include the training program, as well as training aids, simulators, and support equipment used at each level of maintenance. Particular attention always should be given to performance of critical tasks. In addition, tasks that are new, unique, or hazardous must be included in the evaluation, or some assurance should be given that these tasks can be performed satisfactorily.

A training deficiency exists when the training provided does not address the skills needed to operate or maintain the system. Once a deficiency has been identified, an assessment should be made that classifies the deficiency as a shortcoming in the training provided, in the documentation, or in the system itself.

PARAMETERS

Training effectiveness is based on both the training programs and the performance of the individuals while accomplishing tasks associated with the use, operation, and support of the system, to include individual and collective training.

The operational evaluation addresses how well the trained individuals perform the required tasks. The ability to perform the necessary tasks correctly, once the individual(s) is in an operating environment, establishes that the system can be operated and maintained by the personnel so trained. However, criteria are needed for operator and maintainer performance during combat also. These criteria may not be the same as those used during peacetime. For example, changing a tank engine in a maintenance bay is much different from changing the engine in the field, at night, and under rainy conditions.

The parameter "critical tasks demonstrated" is the ratio of critical tasks demonstrated by the trainee using validated procedures within the time standard, to the total number of tasks attempted, or total tasks within the manuals. It can be calculated for each maintenance level or for each skill category (MOS, AFSC, etc.). This parameter has a close alignment with some of the parameters used in the evaluation of documentation (see section 2.8).

KEY POINTS

OT experience can be used to modify the training requirements.

During OT&E, operators and maintenance personnel gain experience with the system and task procedures and skills required to operate and maintain the system. This experience may indicate required changes to the training and skill needs. The evaluation report should indicate if training and skill needs should be changed. Along with these changes, training aids, simulators, and support equipment must be assessed to ensure that they are adequate to support the operational requirements.

OT planning must address when the training program will be available for evaluation.

The delivery of all required training materials and equipment may not coincide with the OT&E schedule. Training manuals usually are not developed on schedule, or they are in an early draft stage. Further, software changes just prior to test may cause major changes in system operations, or personnel available to the test program may have training that is not representative of that planned for the operating units. Unless OT planning considers these possibilities, the OT&E may be unable to evaluate the training program.

The interrelationship between training, documentation (section 2.8), and human factors (section 2.13) must be recognized during the OT planning.

Just as the documentation evaluation is directed at examining various important operating and maintenance tasks, the ability of the personnel to perform the tasks with the documentation provided is dependent on the training those personnel have received. Similarly, human factors and training are related. Tasks that incorporate unusual or complex human factor aspects may require additional or more extensive training. Combined evaluation of these areas may prove to be very beneficial and should be considered during the planning for the OT&E. The adequacy of the training for very complex or demanding tasks should be assessed by reviewing the personnel performance during the actual system operation.

Training and OT tasks should be correlated.

The correlations between the tasks included in the training and the tasks performed during the operational test should be analyzed. Tasks should be identified where the personnel either were not trained or were inadequately trained. There also may be tasks where training was determined to be unnecessary.

Any awkward or unusually demanding tasks that caused personnel problems should be identified.

In some instances, the training requirements and the training planned are based upon an inaccurate view of the system's operations or maintenance activities. OT can identify unusually critical tasks that need to have the training re-evaluated. This activity is closely related to human factors (section 2.13).

"Wartime usage rates" for a system is defined as

the quantitative statement of the projected manner in which the system is to be used in its intended wartime environment.

Wartime usage rates can be expressed in parameters such as flying hours per month, miles per day, rounds per day, or hours per month. The full meaning of these parameters requires a definition of the rate in relation to the planned operational scenario. For example, miles per day has limited meaning unless the "miles" are characterized by speed, terrain, system activity, etc. Similarly, sorties per day is not a valid measure unless the characteristics of the sortie are defined, e.g., mission, weapons carried, sortie length, speed.

The addition of "wartime usage rates" into the list of suitability issues resulted from a concern that, in some instances, the suitability of systems was being assessed at an unrealistic tempo during the operational testing period. "Wartime usage rates" was added to emphasize the need to have suitability evaluations conducted at usage rates that approximate those anticipated during wartime use. Within these frequency measures, there also needs to be a description of the missions themselves, including mission duration. Early in the system's development, these mission profiles must be identified by the acquisition organization in conjunction with the using and supporting organizations. As the system progresses through the development cycle, it becomes more and more important that the usage rates be defined in greater detail. The design must be made in the context of this usage rate, and the logistics support must be planned in consideration of these rates. When OT&E is planned and conducted, the wartime usage rates are a fundamental part of determining how to structure a test to determine if the system can meet its wartime demands.

All of the operational suitability areas contribute to the ability of the system to realize the wartime rates of usage. The number of flying hours, miles, or rounds that the system is capable of providing is dependent on its availability and the balance between the logistics demands and the logistics resources that are provided.

PARAMETERS

To assess the ability of the planned logistics system to support a new weapon system requires that the logistics support be examined in the context of the projected wartime usage rates. When a need is postulated, there may not be an accurate projection of what the wartime usage rate is. These parameters must be identified in the requirements documentation. Example measures are sorties per day for a tactical aircraft, hours per day for simulators and some communications systems, message units per time period for other communications systems, rounds per day for ground combat systems, etc.

KEY POINTS

The usage parameters should be identified early in the program's development.

The parameters or measures for wartime usage should be known and agreed to among the participants in the development process at Milestone I. Usually the parameters for certain classes of systems are relatively easy to specify and to reach agreement on. Examples might be miles per day for a tank or ground vehicle, flying hours per month or sorties per day for aircraft, hours per month for surveillance systems, or rounds per day for an artillery piece.

Usage parameters must be fully defined.

The wartime usage rate must have full definition of the wartime mission or scenario for the rate to have meaning. Message rate per day may be an acceptable usage rate for a communications system, but the content and complexity of the messages must be defined to give the statement meaning. These definitions can be a composite of types, i.e., "x" percent of these message, "y" percent of those, etc. These defined usage rates will be documented in the Operational Mode Summary, Mission Profile, or similar documents.

Usage rates should be developed with the new system's capabilities in mind.

If the system's usage rates are based upon predecessor systems' experience, these rates should be analyzed by the intended users to assure that they still are valid. In many cases, new systems with improved capabilities have been found to have very different use patterns once they are placed with the operating units. Examples are surveillance systems that were used infrequently until they were replaced by higher capability systems. The using organizations were so pleased with the improved systems that the usage increased significantly to almost continuous use.

The operating tempo during OT should be developed from the planned usage rates.

By Milestone II, the value for the wartime usage rates should be known and documented in an Operational Mode Summary, or similar document. These rates also should be included in the Milestone II TEMP, by reference. The developing agencies should understand the usage rates. They also should be used by the agencies performing the logistics planning, and should be part of the requirements used in the planning of the OT&E program.

The OT may be incapable of directly demonstrating the wartime usage rates.

The planned OT should test the wartime usage as specified in the Operational Mode Summary, or mission profile. Any attempt to avoid testing under typical combat conditions/wartime usage rates should be examined carefully, as such unrealistic operation could significantly alter the suitability performance seen during testing and serve as a severe test limitation. If the planned OT is incapable of testing the system at the high level of wartime usage, then modeling and simulation should be used to project the system's capability at rates higher than those seen during the operational testing. This modeling and simulation requirement should be identified by the OTA in the Milestone II TEMP, and planning initiated to develop and validate the required models and simulations.

Some evaluation must be made of the system's capability to perform at the planned wartime usage rates.

The capability of the system to perform at wartime usage rates should be assessed in the OT&E prior to Milestone III, or IIIA if there is one. If there is doubt about the ability of the logistics system to support the system at this rate or the ability of the system to perform at this rate, then the test report should highlight these conclusions. If there is a perceived limitation on the system usage rate, this limitation should be highlighted in the OT&E report. As an example, an aircraft system was incapable of achieving its squadron level sortie rate because of the demand placed on the second level test equipment and the number of test stations planned for each squadron. The highlighting of this deficiency resulted in a re-evaluation of the number of test stations.

2.12 SAFETY

Safety is defined as

freedom from those conditions that can cause death, injury, occupational illness, damage to or loss of equipment or property, or damage to the environment.

Safety is an essential and integral part of assessing a system in an operational environment. It addresses any potential hazards that the system (both hardware and software) poses to personnel, or other systems or equipment. Safety usually is evaluated by observing the system's use and maintenance while performing other portions of the operational testing. Because the safety assessment is a byproduct of this testing, it is important to ensure that safety aspects of the system's use are not overlooked as the principal attention is focused on other aspects of the system's testing. Since the OT&E may be the first instance where the system will be operated and used in its planned environment, this also may be the first instance where it will be possible to observe any potential safety problems.

The acquisition organization usually will conduct system safety programs on more complex systems. The system safety program uses engineering and management techniques to identify and eliminate potential hazards and reduce associated risks.

PARAMETERS

During some operational tests, the OTA may use the categories and hazard levels from the system safety program as a way of identifying the results of the OT from a safety standpoint. The parameters for system safety relate to the number of hazards in specified categories and the projected frequency of exposure to these classes of hazards.

Number of Hazards by Category: MIL-STD-882 has a series of four hazard categories, with Category I, Catastrophic, being the most serious. The categories are:

Description	Category	Mishap Definition
Catastrophic	I	Death, or system loss
Critical	II	Severe injury, severe occupational illness, or major system damage
Marginal	III	Minor injury, minor occupational illness, or minor system damage
Negligible	IV	Less than minor injury, occupational illness, or system damage

For most systems, the objective is to eliminate all Category I and II hazards. Also, the hazards that are identified during operational testing should be categorized, if possible.

Hazard Probability: If possible, any observed hazards should be identified by the probability levels that are used in system safety programs. This designation may aid in the investigation and resolution of the hazards. The hazard probability levels (contained in MIL-STD-882) are:

Level	Probability	Definition
A	Frequent	Likely to occur frequently
B	Probable	Will occur several times in the life of the item
C	Occasional	Likely to occur sometime in life of an item
D	Remote	Unlikely, but possible to occur in life of an item
E	Improbable	So unlikely that it can be assumed occurrence may not be experienced

KEY POINTS

Operational testing provides an opportunity to observe the system operated and supported by personnel having the expected skill levels.

Prior to operational testing, the personnel who operate and maintain the system probably will have higher skill and experience levels than will the planned operational personnel. Therefore, the OT is the first opportunity to observe the system in the hands of personnel with the projected levels of experience and skills. This first observation may indicate potential safety problems that were not observed in the earlier testing. Test planning should focus attention on detecting and documenting any new or unexpected hazards to personnel. This observation is closely related to human factors assessment (see section 2.13).

Observers of operational tests should be sensitive to any potential for significant hazards.

All personnel who are involved in or associated with the conduct of the operational testing have a responsibility to identify any potential hazard, and cause the test to be stopped if a hazard in Categories I or II is perceived. In most cases, operational testing should be conducted without outside interference, but safety is an exception.

Safety testing should consider the operating environment of the system.

Any safety-oriented testing or assessment should consider the entire expected range of environments. Some safety features may be very effective in good weather on a clear day. Hazards may be clearly seen and easily avoided. In poor lighting or in bad weather, poor visibility may result in unexpected hazardous conditions.

Software faults can result in unexpected hazards.

Faults observed in the software should be evaluated for potential contribution to hazardous conditions. As an example, for aircraft systems, software faults can impact flight safety.

2.13 HUMAN FACTORS

The term "Human factors" is defined as

those elements of system operation and maintenance that influence the efficiency with which people can use systems to accomplish the operational mission. The important elements of human factors are the equipment (e.g., arrangement of controls and displays), the work environment (e.g., room layout, noise level, temperature, lighting, etc.), the task (e.g., length and complexity of operating procedures), and personnel (e.g., capabilities of operators and maintainers).

This suitability element addresses the compatibility between system hardware and software elements and the human elements. It is intended to identify system performance problems, human task performance problems, and hazards to personnel under realistic conditions of combat use. While there is a close alliance between the examination of human factors and the examination of manpower supportability (see section 2.9), training (see section 2.10), and safety (see section 2.12), "human factors" is focused more on the hardware and software elements of the system. It is more of an evaluation of the system itself, what the system requires of the people who operate and maintain it, and how the system fits into the relationship with the people who are going to operate and maintain it.

Evaluation of the human-factor aspects of a new system includes all of the interfaces between personnel and the hardware and software. It includes also the interfaces with both the system operators and the system maintenance personnel.

The human-factor considerations include compatibility of the man-machine interface. Considerations in this area comprise information displays (machine feedback), symbology (standard versus unique), operator controls, personnel comfort and convenience, portability of the equipment (bulk, weight, load distribution, straps, handles, etc.), accessibility for operation and maintenance, physical workload demands, mental workload/information processing demands, compatibility with task characteristics, task environment, etc.

Traditional human-factor testing may address also ergonomic (e.g., can the operator see or reach the indicator or control) considerations, although this area normally is part of the developmental testing. The OT&E should address the operator's effectiveness and efficiency in the performance of assigned tasks.

PARAMETERS

Human factors are evaluated qualitatively using checklists that focus attention on the human-factor aspects of the system. They also can be evaluated quantitatively by performing timed tasks. Data to assist in the evaluation may be collected on task times, response times, error rates, accuracy, etc. Interviews, questionnaires, and debriefings of operators and maintenance personnel can be used to gather data on impressions of displays, man-machine interface, accessibility, portability, task environment, task difficulty, unnecessary steps, work space, personnel fatigue, etc.

KEY POINTS

Human factors should address both operators and maintenance personnel.

Human-factor evaluation should address the ability of the operators to use and to control the functioning of the system. It also should ensure that the support personnel have the access and the physical capability to perform the required maintenance.

The software interface with personnel should be assessed.

As systems become more software intensive, there is a need to evaluate the interface of the software with the operators and/or maintenance personnel. How does the software present information? Is it clear, or can there be misinterpretation? Is there consistency among the various software displays so the operators will have an acceptable learning curve as the system is used? Is software designed with safeguards, to the extent possible, against system failures due to incorrect key entries? Some OTAs use checklists and questionnaires to examine the "usability" of the software.

Physical demands on personnel should be assessed.

Consideration needs to be given to the manner in which the system is to be used under combat conditions. Are the personnel likely to be operating or maintaining the system while wearing protective clothing against the weather or against chemical, nuclear, or biological attack? How long will they have to operate under these conditions for the system to perform its mission(s)? Does the system require relatively long periods of concentration, or exertion? Some soldier-fired systems require the sights to be kept on the target for a long period of time. Is this period realistic in terms of human capability for the average person?

The employment of new or advanced display techniques should be identified.

When new or advanced display techniques are used in the operator's station or in maintenance equipment, there may be significant questions about how these changes will be accepted and integrated into the operation of the line operating unit. A COI should be identified and the OT should ensure that any potential problems are examined in adequate detail.

Human factor conditions should consider the entire operating environment.

The examination of the man-machine interface needs to consider that the system will be operated under a wide rage of conditions. If needed, can the system be used effectively with arctic clothing, CBR protective clothing, etc.? Can it be effectively used in poor weather, or with limited lighting? In one example, the OTA took user troops to the contractor's facility where they found that the systems could not be assembled by the planned field personnel. They also discovered that tactical personnel could not enter some of the shelters with packs and weapons, forcing them to leave these items outside in potentially hostile or contaminated environments.

Combat stress conditions can affect the ability of personnel to operate or maintain the system, and should be evaluated.

Concerns for the ability of personnel to use and repair the system under combat stress conditions must be addressed. Under very stressful conditions, often only the simplest tasks succeed. Simulating the combat stress factor is very difficult. The number and frequency of different tasks should be analyzed and documented in the operational test plan. The plan should include the measures that will be used to gain insight into the reactions of the people to the stress of the proposed environment, and their ability to perform as required.

Chapter 3

OTHER OPERATIONAL SUITABILITY ISSUES

Besides the suitability elements identified in the definition of operational suitability and discussed in Chapter 2, other issues are essential to a discussion of operational suitability in OT&E. For operational suitability to be evaluated effectively during OT&E, the following issues must be addressed and emphasized: suitability modeling and simulation, integrated diagnostics, environmental factors, electromagnetic environmental effects, and software supportability. These issues must be understood and examined in light of their relationship to the suitability elements discussed in Chapter 2. Those relationships are discussed briefly in the paragraphs that follow.

Suitability modeling and simulation (M&S) has the potential to aid in the operational evaluation of suitability by supplementing the actual operational testing. Modeling and simulation can be very valuable in identifying and focusing the test and evaluation effort. If structured properly, the M&S can provide analysis results before the testing is planned to show which of the suitability areas is most critical to the successful employment of the system and which are most critical in the planning of the testing activity. Once the testing is completed, the M&S can aid in the evaluation of the test data by expanding the framework of the actual testing and providing insight into the meaning of the data when examined in light of other operating scenarios or by simulating other combinations of events. The risk in the use of the M&S is that it will become used in ways that are not totally correct. Improper or limited M&S can result in the decrease the importance of actual operational testing. M&S is not a substitute for properly planned and conducted testing.

Integrated diagnostics is a structured process which maximizes the effectiveness of diagnostics by providing a cost-effective capability to detect and unambiguously isolate all faults known or expected to occur in weapons systems and equipment in order to satisfy weapon system mission requirements. This capability is extremely significant during wartime when it is imperative that critical failures be found and fixed quickly.

Environmental factors, as shown in Table 3-1 (page 49), affect all of the suitability elements. Items that may be adequate in good weather and lighting may be totally unacceptable in restricted visibility. The range of terrain and environments that vehicles must overcome is another important factor. An acceptable operational test must ensure that the environmental limitations that exist do not invalidate the test results when applied to the system's intended operational location.

The issue of electromagnetic environmental effects (E3) which is included as a special emphasis item with the area of compatibility of the system E3, requires special techniques and knowledge. As the modern systems incorporate a higher density of electronics, the importance of evaluating and understanding the E3 area deserves special consideration.

Software is a similar area. The considerable amount of software included in a modern system increases its importance and its criticality to the successful use and operation of the system. Software can be viewed as having reliability, maintainability, and availability impacts. Likewise, system software should be part of the evaluation of training, documentation, and human factors. Conducting the proper OT in these areas depends on the ability of the persons involved in planning and conducting the OT and their familiarity with software evaluation methods.

3.1 SUITABILITY MODELING AND SIMULATION

The DoD is in the process of issuing expanded guidance on the development, validation, and use of modeling and simulation (M&S) in the acquisition process. In January 1989, the Director of Operational Test and Evaluation (DOT&E) issued the "DOT&E Policy for the Application of Modeling and Simulation in Support of Operational Test and Evaluation."

A model is defined as

> *a representation of an actual or conceptual system that involves mathematics, logical expressions, or computer simulations that can be used to predict how the system might perform or survive under various conditions or in a range of hostile environments.*

Simulation is defined as

> *a method for implementing a model. It is the process of conducting experiments with a model for the purpose of understanding the behavior of the system modeled under selected conditions or of evaluating various strategies for the operation of the system within the limits imposed by developmental or operational criteria.*

There are several different types of simulations, including those that use analog or digital devices, laboratory models, or "test-bed" sites.

The use of properly validated M&S is strongly encouraged during the early phases of a program to assess those areas that cannot be directly observed through testing. The use of M&S is not a substitute for actual testing; however, it can provide early projections and reduce test costs by supplementing actual test data.

The use of modeling and simulation in the operational suitability area can provide a number of benefits. M&S can be used to focus limited test resources by identifying the critical elements in a logistics support system, e.g., the choke points for the flow of the support resources. M&S also can be used to translate the rate of use in the test scenario to the wartime usage rate. If, for example, test aircraft are flying only one or two sorties per day, the "load" on the support resources is significantly different than if a higher, wartime sortie rate were being flown. M&S can aid in assessing the impact of these differences. M&S may be also used to evaluate elements of the support system that are not present at the test site. For example, if the second-level maintenance capability (test equipment, facilities, etc.) is not available, then a properly constructed and validated model can be used to provide insight into the ability of the planned second-level maintenance facility to support the system.

PARAMETERS

The key parameters in M&S are the assumptions and ground rules used in inputting data. The output can be valid only if valid assumptions and ground rules have been used.

KEY POINTS

Plans for M&S should be evaluated for potential credibility of the results.

The credibility of the results of M&S is a judgment formed from the composite of impressions of the inputs, processes, outputs, conclusions, the persons or agencies involved, and the strength of the evidence presented. Appendix B of the "DOT&E Policy for the Application of Modeling and Simulation in Support of Operational Test and Evaluation" provides a series of questions to assist in assessing M&S results' credibility. These questions provide a good outline for examining M&S activities.

All of the models planned for use on the OT&E program should be accredited for the purpose.

Accreditation is defined as the process of certifying that a computer model has achieved an established standard such that it can be applied for a specific purpose. This means that management has examined the model and, based upon experience and expert judgment, has declared that the model is adequate for its intended use.

Detailed definitions of planned operating and support scenarios are essential for a valid M&S effort.

In many cases, the detailed definition that is needed for M&S is beyond that existing in program documentation. This is particularly true in the suitability area, where the maintenance and supply concepts to be used must be defined in detail. There is a potential for the M&S results to be driven by some of the necessary assumptions rather than by the system characteristics. On the other hand, if the responsible personnel and organizations are requested to provide the required detail, and the support planning is thought through, then other organizations within the program also will benefit.

The latest program information must be incorporated into the M&S activity.

Many modeling efforts lack current information. Program conditions may change. The system design may be revised, or new threat information received. In each case, the earlier model may be invalidated. Assuring that the modeling results reflect the best and most current information available is an important consideration. Procedures must be established to assure that current information is provided to those doing the modeling and evaluation of the simulation results.

Defined plans for the use of M&S should be presented in the TEMP.

The TEMP should indicate any plans for the use of suitability modeling and simulation to complement or supplement the operational testing. The models to be used should be identified, and plans for their validation described. M&S should not be used in place of actual testing.

The TEMP and other test documentation should include a discussion of the rationale for the selection of the specific models that are planned for suitability analysis.

Models are used for many of the assessments for suitability. The Services should list the models and discuss their advantages/disadvantages in the TEMP or other test documentation so that an evaluation can be made as to the utility of the model. The DOT&E evaluator must be able to assess the validity of the selected models and of the OTA's assessment results from the model's use.

3.2 INTEGRATED DIAGNOSTICS

The term "Diagnostics" is defined in the OTAs Multi-Service Testing MOA as

the ability of integrated diagnostics (automated, semi-automated, and manual techniques taken as a whole) to fault-detect and fault-isolate in a timely manner.

"Integrated Diagnostics" is defined as

a structured process, which maximizes the effectiveness of diagnostics by integrating pertinent elements, such as testability, automatic and manual testing, training, maintenance aiding, and technical information that will satisfy weapon system peacetime and combat mission requirements and enable critical failures to be fixed with minimum loss of operational availability.

The purpose of integrated diagnostics is to provide a cost-effective capability to detect and unambiguously isolate all faults known or expected to occur in weapons systems and equipment in order to satisfy weapon system mission requirements. In wartime, this becomes extremely significant in that it is imperative that critical failures be found and fixed quickly to support combat turn-around times, which can equalize battles against numerically superior forces

The term "Diagnostics" often is used as a general term to cover all means of determining that a system fault has occurred, and the means to determine where the fault is and to isolate it to a portion of the system that can be repaired or replaced. There are many other terms that are used in this area, including Built-In Test (BIT), Built-In Test Equipment (BITE), Built-In Test and Fault Isolation Test (BIT/FIT), and Automatic Test Equipment (ATE).

The key to integrated diagnostics is the successful consideration and integration of the functions of detection, isolation, verification, recovery, recording, and reporting, in a comprehensive and cohesive fashion, with the operator and with support functions that may be automatically, semi-automatically, and/or manually controlled.

PARAMETERS

Two parameters are listed in the OTAs MOA for diagnostics use in multi-Service OT&E test programs. They are:

Percent of correct detection given that a fault has occurred (P_{cd}),

$$P_{cd} = \frac{\text{The number of correct detections}}{\text{The total number of confirmed faults}}$$

Mean Time to Fault Locate (MTTFL),

$$MTTFL = \frac{\text{The amount of time required to locate faults}}{\text{The total number of faults}}$$

These parameters apply to a specified level of maintenance and, therefore, may be applicable to each level of maintenance.

The ability of an automated diagnostics system to isolate faults also may be measured. One parameter for this characteristics is percent fault isolation.

$$\text{Percent Fault Isolation} = \frac{\text{Number of fault isolations in which automated diagnostics effectively contributed}}{\text{Number of confirmed failures detected via all methods}} \times 100$$

Another important measure of the capability of the diagnostics system is to identify how frequently the automated diagnostics indicates that a fault exists when in fact the system is functional. This area is particularly troublesome, since the system false alarms may be either improper indications of faults that do not exist, or faults that did exist but were transient in nature. Identifying which situation exists is most difficult for complex systems. One of the more common parameters for false alarms measurement is BIT false alarm rate (expressed as a percentage).

$$\text{Percent BIT False Alarm} = \frac{\text{Number of BIT indications not resulting in maintenance actions}}{\text{Total number of BIT indications}} \times 100$$

Another important aspect of integrated diagnostics is the compatibility of the various levels of testing. The faults that are detected by the automatic built-in test must be verified by a more thorough diagnostics system than is available to maintenance personnel. The verification of a fault after removal of the equipment from the system platform is sometimes exasperating because the maintenance test station cannot duplicate the operational environment (vibration, temperature, etc.) that was present in the instance of the initial fault reporting. Once the failed item is removed, the test equipment at the next level of maintenance must be able to identify the same fault. Two parameters are commonly used in this area: the "cannot duplicate" (CND) rate and the "retest okay" rate. Both are expressed as a percentage.

$$\text{Percent Cannot Duplicate} = \frac{\text{Number of faults that could not be duplicated by later maintenance actions}}{\text{Total number of faults reported}} \times 100$$

$$\text{Percent Retest Okay} = \frac{\text{Number of faults that could not be duplicated at the next level of maintenance}}{\text{Total number of faults reported}} \times 100$$

These last three equations are based on a percentage of BIT indications. Improved parameters, used on some new systems, use a measure of life units (hours, miles, sorties) in the denominator. This results in parameters such as number of BIT false alarms per sortie or per hour.

In addition to the parameters discussed here, each of the Services employs numerous other general and unique parameters in their respective programs. These include parameters that relate to the manual, as well as automatic and semi-automatic aspects of integrated diagnostics.

3.2 INTEGRATED DIAGNOSTICS (Cont'd)

KEY POINTS

The approach to system diagnostics should be discussed in the early system planning documents.

These discussions provide a basis for relating the diagnostics requirements to other system parameters, such as reliability, maintainability, and availability.

All aspects of the integrated diagnostics function must be planned for.

All integrated diagnostics test items required in the support of the weapon system must be planned for. Focusing on the exotic on-board built-in test features can lead to only minimal planning for the less exotic support functions such as automatic test equipment, test program sets, technical manuals, training, and required skill levels of personnel.

The program manager should have firm diagnostics requirements established before Milestone II.

Diagnostics requirements should be the result of analysis that allocates the diagnostics requirements across the various alternatives, i.e., automated systems, semi-automated systems, test equipment, and manual troubleshooting. The initial operational testing should provide insight into the system's capability versus these allocated levels of diagnostics. The total diagnostics capability may meet the threshold, but if some elements are far from what is required, the system still may be not suitable.

Diagnostics short-falls should be evaluated by the OTA as to the total impact on the system and its support resources.

When operational test results are presented at the Milestone III decision, the evaluation of diagnostics capability should discuss the relative effect of the diagnostics performance on the reliability, maintainability, and availability of the system. For example, what is the impact on the system availability if the P_{cd} is less than the threshold?

Diagnostics short-falls may be obscured by activities in other suitability areas.

While many current systems have failed to realize the level of required diagnostics, these deficiencies have not always been corrected prior to the system being fielded, but have been offset by changes in other parts of the support system. For example, the inability of automated systems to perform the level of fault isolation that was expected may lead to an expanded use of manual troubleshooting on some portion of the system. The allocation of the diagnostic task to the various alternative methods must be assessed as part of OT&E. If the allocation cannot be realized, then the impact of a reallocation must be assessed, along with the penalties in cost or readiness that an adjustment of the allocation conveys.

Indications of poor performance of the system on-board diagnostics early in the program should be followed closely, as lack of diagnostics performance can lead to major suitability problems.

Failure of a system to perform the planned-for on-board diagnostics can have serious impacts on the support structure. If it becomes necessary to perform those functions off-platform, adequate types and quantities of support equipment may not be planned for and the planned training and personnel skill levels may not be able to absorb the required additional burden.

A common problem with diagnostics is its immaturity at the early stages of operational testing.

When the system is tested in its early stages, the diagnostics may be less capable than desired, or result in numerous false fault indications. Resulting deficiencies are labeled as the result of immaturity and, more often than not, it is projected that the mature system will not have these problems. The maturing of a diagnostics system is a difficult and demanding task. Revising the diagnostics approach to a system design that is fairly fixed generally will not yield significant improvement. The expectation of significant improvement must be accompanied by a maturation program that has the proper resources to do the job and the operational testing to verify the results.

The automated diagnostics capability of a system usually improves as the system's design matures.

The automated diagnostics capability of a system is one of the system features that usually is not completed with the initial design. The testing and operation of the system provides additional insights into both the system's performance and the potential failure modes and effects. Such information that results from development activities should be used to improve the diagnostics capability. The impact of this situation on OT&E is that the maturity of the diagnostics at the time of operational testing needs to be evaluated prior to the testing, and the test results need to be evaluated in light of system maturity.

Poor diagnostics performance can have serious effects on the system's suitability.

If the system has poor diagnostics performance during operational testing, the impact may be felt in a number of areas. The operators and maintenance personnel may loose confidence in the diagnostics system. If incorrect system status is frequently displayed to the operators, they will be unable to rely on the system displays (this can be a particular problem with protection systems, such as electronic warfare systems). Similarly, if the maintenance personnel perceive that the automatic diagnostics system is not accurate much of the time, they will have to resort to other means to maintain the system. This may result in unrealistic data from the OT or unnecessary demands being placed on the supply system, or it may cause additional requirements in documentation, training, or other logistics areas for the operational system.

3.3 ENVIRONMENTAL FACTORS

The "operational environment" is a critical factor to the operational suitability of a system. The ability of the DoD operational testing organizations to determine the operational suitability during OT&E is dependent on how the test environment compares to the operating environment. The "operational environment" is composed of many individual, distinct and almost unrelated areas. Often, each area must be addressed separately to ensure adequate, deliberate consideration.

In the context of this guide, the definition of "environment" maintains a broad scope and includes the weather; vegetation; terrain (land or water); acoustic; electrical/electronic; illumination; chemical, biological, radiation (CBR); and battlefield conditions. There are two major categories of environment, natural and man-made.

The framework for discussing environments is shown in Table 3-1.

Any environmental condition may have an impact on the system's performance and the ability to properly use the system in the intended combat or wartime environments. The word "environment" also may be modified by an appropriate explanatory adjective, e.g., combat environment, human environment, vibrational (mechanical) environment, and so forth. Taken together, these encompass what might be considered the "operational environment." Care must be exercised in the preparation of OT&E documents to ensure that the writer and the reader similarly interpret the discussion of environment.

When an operational need is stated for a new system, it is necessary to state what the conditions for use will be. These use conditions include the environmental factors that bear on the utility of the system. Any system limitation that is postulated due to environmental factors or conditions should be identified by the user or user representatives who are responsible for developing the system level requirements. These limitations also should be identified for examination as part of the OT&E. The OT&E should be planned in such a manner so as to determine if the level of limitation is as expected, or if it is more severe than estimated. Testing also should determine if the limitation affects the system in a manner other than what was predicted.

The operational requirement should state the general operating environment and indicate if the requirement includes any limitation to operational use due to the environment. That is, does the system comprise elements that are sensitive to environmental conditions (e.g., rain, fog) and also battlefield conditions (e.g., smoke, dust)?

Table 3-1 Framework for Discussing Environments

ENVIRONMENT		NATURAL (EXAMPLES)	MAN-MADE (EXAMPLES)
WEATHER		Rain, Snow, Winds, Sea State, Fog	——
VEGETATION		Grass, Shrubs, Trees	——
TERRAIN		Swamp, Desert, Mountains, Ice, Plains, Water, Soil	* Moats, Fox Holes, Tank Traps, Roads, Urban Features
ACOUSTIC		Thunder, Rain, Fish, Whales, Waves	* Decoys, Ships
ELECTRICAL/ ELECTRONIC		Lightning, Solar Flares, Ionospheric Disturbances	* Jamming, EMP
ILLUMINATION		Sun, Moon, Eclipse	* Flares, Searchlights
CBR		Space Radiation, Epidemics	* Nuclear Radiation, Germ Warfare, Toxic Gases
B A T T L E F I E L D	SMOKE	Vegetation Fires	Target Hits
	DUST	Dust Storm	Bomb Blast
	DIRT, SAND	Sand Storm	Bomb Blast
	OBSCURANTS	Clouds, Rain, Fog, Snow, Haze, Sand, Dust	* Smoke Canisters, Flares, Battle Dust and Debris

* Enemy actions or countermeasures that impact on survivability or susceptibility are evaluated as components of operational effectiveness and are not addressed under operational suitability.

3.3 ENVIRONMENTAL FACTORS (Cont'd)

PARAMETERS

Environmental parameters can be used to characterize the intended use environment, e.g., terrain that the system is intended to travel over, or they can be used to characterize the system's capability versus the environment, e.g., minimum visibility level at which the seeker will be capable of operating. The first category is used to communicate the user's environmental requirements, while the second communicates the system's capability within the environment. Examples of parameters that communicate the environment are:

Terrain Grade: The incline a ground vehicle should be able climb, given its power and traction

Water Depth: The water level through which a ground vehicle should be able to pass

Sea State: Ocean wave conditions under which a vessel should be able to perform certain mission functions.

Examples of parameters that communicate the system's capability within an environment are:

Range: The detection range of a seeker under certain specified obscurant conditions

Speed: Vehicle speed over specified terrain conditions.

KEY POINTS

Most systems have environmental limitations.

These limitations are defined as the degree of conditions (e.g., weather, sea state) under which the system will not be able to operate effectively. That a system will not be able to perform in extreme adverse conditions is an accepted fact for most systems, but the threshold for non-effectiveness must be known to the user.

Environmental limitations should be quantified and understood.

Early in the acquisition process, the ranges of conditions under which the system must be effective should be clearly established by the user or the user's representative. The frequency of occurrence of adverse conditions (e.g., weather) that will limit system performance must be understood to permit acquisition decisions and appropriate planning by the Service. For each system, available data should be used to quantify the frequency of occurrence of any limiting conditions. The specified range of acceptable environmental conditions and identification of significant limiting factors should provide an important consideration in the Service's planning for the OT&E.

The system requirements documents should discuss the required operating environment.

The requirements document must address the intended operating environment for the system. Under what conditions (weather, terrain, vegetation, etc.) will the system be employed?

Limitations to system operation and/or maintenance should be projected prior to OT.

If there are any expected limitations to either the scope of operations or the system's capability, then these should be documented by the acquisition agency prior to Milestone II. The planning for operational testing should identify any environmental Critical Operational Issues. The OT&E needs to address the environmental conditions such that any limitation on capability due to environmental conditions is either verified or further understood and defined.

Personnel who operate and maintain the system are affected by the environment.

The ability of operators and/or maintenance personnel to function under certain environmental conditions also needs to be addressed. Personnel usually are not stressed to their endurance levels, but the impact of the weather, protective clothing, and reduced visibility may be factors that impact the efficient use of a system. If the system design requires critical personnel interaction, then consideration of these environmental areas should be part of operational test planning.

Systems with optical sensors can have limited performance in some environments.

Systems that have electro-optical, infrared, or millimeter wave seekers, or that require visual sighting by the operators, may have limitations when used in areas with high levels of obscurants, e.g., smoke, dust, etc. Any system limitations that are postulated need to be estimated and included in program planning information. The level of the limitations should be identified as a Critical Operational Issue if the limitation is critical to the eventual use of the system in its intended environment.

Environmental conditions at the OT sites usually are limited.

The sites for operational testing usually are limited by available funding. Normally there will be only one site for the early operational testing and this site is more likely to be selected for instrumentation, test facilities or ranges, or test organizations than for weather, terrain, or vegetation conditions that are representative of the intended operational environment. System requirements should clearly state if different or additional environmental conditions are important to understanding the system's operational effectiveness or suitability. For example, terrain testing is usually part of DT; if specific problems are expected with terrain or vegetation, then short operational test phases or demonstrations should be considered to address these areas.

Operational testing usually is not performed outside the system's intended environmental operating envelope.

The planning of the operational test program should address the intended operating environment and generally should not incorporate plans to operate the system outside that environment.

Operational testing may determine additional environmental limitations.

If the potential for additional limitations is great, then OT&E must also attempt to define any additional environmental limitations on the system's performance. For example, a system may be found to be not maintainable under chemical attack. If the item must be decontaminated prior to some maintenance tasks, and the capability to do this does not exist, the system has an obvious major limitation. At Milestone III, test results should be adequate to verify the level of any environmental limitation.

3.4 ELECTROMAGNETIC ENVIRONMENTAL EFFECTS (E3)

Electromagnetic Environmental Effects (E3) is defined as

the impact of the electromagnetic environment upon the operational capability of military forces, equipment, system, and platforms. It encompasses all electromagnetic disciplines, including electromagnetic compatibility/electromagnetic interference; electromagnetic vulnerability; electromagnetic pulse; electronic counter-countermeasures; hazards of electromagnetic radiation to personnel, ordnance, and volatile materials; and natural phenomena effects of lightning and precipitation-static.

Compatibility with the electromagnetic environment is an important issue in the system's overall compatibility. E3 includes the subjects of electromagnetic interference (EMI) and electromagnetic compatibility (EMC). Within the operational suitability area, these subjects are examined as they relate to companion or friendly systems. Vulnerabilities to enemy electronic systems are addressed under operational effectiveness. To properly assess the E3 area requires the consideration of many unusual situations that may cause incompatibilities within the E3 areas. Understanding these situations requires experience and knowledge of system operation and system use in the intended operational environment. Are there unforeseen items in the environment that will cause problems with the E3 conditions? What companion systems need to be considered? Are there unusual situations in the system's use that will place it with other systems that have an E3 consideration?

E3 addresses the extent to which a system's performance is degraded by electromagnetic effects due to its proximity to another electronic system. EMC and EMI are evaluated for their impact on the electromagnetic transmissions of multiple interfacing systems, as well as for their impact on friendly systems for which interfacing is not intended. Specifically, if two systems have electrical transmissions and are not integrated, but are brought into proximity in operational use, then consideration of their mutual operation in the presence of each other is a part of EMC. EMI addresses interference of components within the same system.

PARAMETERS

While discrete engineering parameters (e.g., spurious emission levels, radiation leakage, interference-to-noise ratios) are quantitative and measurable in the development environment, the focus during Operational Testing is at a higher total system and environment view. Parameters must be focused more on the external relationships, as well as the internal relationships. For example, Operational Testing might verify that vastly different systems and/or multiple copies of the same system can operate suitably while in close quarters at a common site. The objective of the test would be to determine if co-site interference problems exist, which would require the operator to turn one system off when using another system.

KEY POINTS

When viewed as technical considerations, E3 problems can be overcome.

The principal risk area in E3 is associated with overlooking some potential E3 condition, and then not discovering the problem until late in the development process, or until the system is fielded. DT will examine many aspects of E3. The role of operational testing is to provide a realistic E3 environment and thereby identify any potential problems that were not identified earlier in the system's development.

Susceptibility to enemy systems usually is adequately evaluated under operational effectiveness. Compatibility with friendly systems is not adequately addressed.

Adequate attention generally is given to assuring that the system being developed is not susceptible to interference from enemy or foreign equipment. Attention also must be given to compatibility with friendly systems. Examples include

- other systems that are employed by the user organization or the same military Service and are placed in proximity to the system being developed. For example, other Army systems used near the Army system under development.

- other systems that may be used in proximity to the system being developed by other military Services. For example, are there USMC or Air Force ground electronics systems that will be used in proximity to an Army ground electronics system that is being developed?

- other systems under development. Compatibility may be examined during OT&E with the systems that are existing in the operating units, but are there other systems in development that will be major factors in the E3 environment of the new system. Are these items included in the planned operational testing?

The operational testing environment needs to represent the total E3 environment to the maximum extent possible.

Complementary systems and unusual conditions need to be included in the E3 assessment.

Situations should be identified where systems are used to complement each other in ways that are not considered the norm, but which are part of the expected system capability. Joint operations of systems by two or more of the military Services are likely to introduce situations that need to be examined. Other E3 conditions may result from operations in unusual environmental conditions (e.g., weather, terrain). Surface ships operating in high sea states may have E3 environments that are different from those anticipated because of the ships attitude at various times. These conditions need to be examined and considered during the planning for OT&E.

Friendly compatible systems must be identified.

The criticality of these systems must be known so that limited test resources can be focused on examining the E3 environment with these items. Does the test plan list the friendly systems to be included in the test, as well as the systems that are not? At Milestone IIIA, there will be an initial assessment of the E3 risk of the system. An early assessment may be possible by examining available E3 area technical test data. At Milestone III, the compatibility with the friendly systems should be known and the risks areas identified. The operational testing should have included all of the systems that were planned to be part of the testing.

3.5 SOFTWARE SUPPORTABILITY

For the purposes of this guide, software supportability is defined as

a measure of the adequacy of products, resources, and procedures that are needed to support the software component of a system.

Software support activities are necessary to establish an operational baseline, install the software in the system, modify or update the software, and meet the users' requirements. Software supportability is a function of the quality of the software products themselves, the capabilities of the software support resources, and the adequacy of the life cycle processes that affect the procurement, development, modification, and operational support of the software.

The criticality of the software supportability is best exemplified by the need for some system software to be periodically revised or updated to correspond to new situations. Electronic warfare systems are periodically updated as new information on threats is received or new tactics are implemented. The ability to revise the software in a timely and efficient manner can be critical to the suitability of the system to perform its required mission.

PARAMETERS

The methods for assessing the suitability of system software have evolved over the last ten years. Most of the activity by the OTAs in this area has resulted in qualitative evaluation methods using questionnaires, with the results being converted by scoring methods into quantitative measures. For example, maintainability evaluation of the software for a specific system might be scored as a "C." This means that the average qualitative evaluation of the software against a maintainability evaluation questionnaire resulted in a judgment that the software "generally agreed" with the statements in the maintainability questionnaire. A range of responses is possible: "A" = completely agree, "B" = strongly agree, "C" = generally agree, "D" = generally disagree, "E" = strongly disagree, and "F" = completely disagree.

In some software evaluations, maturity levels also are reported. The maturity levels may be expressed in the "number of software changes" or in "software change points." The term "software change points" is used when the individual software changes that are identified as being needed are weighted by the severity of their operational impact on the system. This multiplication results in a parameter termed "software change points."

KEY POINTS

The software documentation can be the key to the effective support of the software.

The timely delivery of software documentation is a key element in allowing the life cycle support activity to maintain and upgrade the software. Review of the documentation has been done in the past using a checklist approach. This approach was useful when the software was of a limited scope. However, with more complex systems, the amount of software is such that a manual, exhaustive inspection is no longer possible. Sampling techniques or other approaches that are to be used need to be identified in the TEMP and the test plan.

The maintainability of software depends on its design and arrangement.

The characteristics of the software that determine its maintainability include its modularity, descriptiveness of the software code and documentation, consistency throughout the code and documentation, simplicity, expandability (is the code built with the objective of making it easy to expand?), and instrumentation (does the code allow the easy use of testing aids?). These characteristics have been evaluated by the OTAs primarily through the use of review questionnaires. The application of the questionnaires to large software packages, the selection of samples to examine, and the relationship of the sample results to assessment of the entire system software are risk areas that need to be examined during the test planning and execution.

The interface that the software presents can be critical to system operation.

The displays, menus, etc., that the system software creates are at the critical junction between the computer-driven system and its user or maintainer. The principal evaluation methods relate to the reaction of the person to the displays. Does the software present clear and understandable information? Is there a consistency throughout the systems operation, e.g., similar menu usage, key stroking similar in similar situations, complex tasks have easily understood sequence, etc.? The evaluation of the user-software interface may be done by questionnaire or by qualitative assessment at debriefings. The questionnaire method results in more consistent and perhaps more thorough evaluations, but the unstructured reaction of the personnel involved should not be ignored. Dissimilarity with the predecessor system may cause negative reactions.

The ability to maintain and modify the software depends on the adequacy of the software support resources.

Evaluation of the planned software support resources consists of the evaluation of the support personnel, support systems, and facilities that are planned. Evaluation of personnel consists primarily of identifying the number of people and the skill levels needed to provide the required support. The software support system comprises the computers and, in some cases, unique software that is needed to provide software maintenance, modification, and upgrades.

The maturity of the software can be evaluated by examining the faults or errors that have been found and the status of the individual corrective actions.

As the software is tested, errors or faults are found. Some OTAs evaluate each software fault (i.e., weight each fault by the severity of the effect on the system) and produce a plot of the cumulative number changes, or change points, against the amount of testing that has been performed. This curve generally shows an increasing number of faults as the software is exercised, followed by a decreasing rate indicating that most of the faults have been found. The use of severity weighting assures that inordinate weight is not given to a number of minor faults, while major faults are ignored. The plot of software maturity can give a clear view of when the software is reaching a maturity, when it is reasonable to start OT, and when the software is mature enough to enter the operational inventory. The DOT&E staff assistant should be familiar with this software maturity technique if it is to be used on one of the assigned programs.

Software maturity depends on testing exposure.

The assessment of the maturity of the software depends on the thoroughness of the software testing. Are all software capabilities being tested? Are the low probability paths, as well as the nominal conditions, being exercised? Are error routines and fault identification modules being included in the testing? Accurate judgment of software maturity necessitates assurance that all aspects of the software are included in the test.

Chapter 4

TEST AND EVALUATION MASTER PLAN (TEMP)

The TEMP is an essential test and evaluation (T&E) document used by the Office of the Secretary of Defense (OSD) to support milestone decisions by the Defense Acquisition Board (DAB). The TEMP is the basic planning document for all T&E activity related to a particular system acquisition. It defines both Developmental Test and Evaluation (DT&E) and Operational Test and Evaluation (OT&E) associated with system development and acquisition decisions. The TEMP relates program structure, decision milestones, test management structure, and required resources to critical operational issues, critical technical issues, evaluation criteria and procedures.

One of the more significant functions of the TEMP is to document test and evaluation issues and criteria that will be considered in acquisition decisions. Thus, the reviewer of a TEMP must realize that the TEMP serves not only as a major control mechanism, but also to provide a clear correlation between issues and program objectives through test-verifiable criteria. In reviewing a TEMP in the area of suitability, one must ensure that it contains pertinent suitability-related information on the system's required operational and technical characteristics, test objectives, and the evaluation process.

Suitability-related system requirements, program structure, technical and operational characteristics, and associated thresholds in the TEMP must be reviewed to ensure that they are consistent with the Requirements Documents, Acquisition Decision Memorandum (ADM), and approved System Concept Paper (SCP)/Decision Coordinating Paper (DCP). T&E must be defined sufficiently to ensure that the test program will assess the effects of human performance on the weapon system's ability to meet all suitability standards, including reliability and maintainability.

The TEMP must include the system's suitability-related critical technical and operational issues and thresholds and their relationship to the system's requirements. It should clearly outline the planned T&E process.

A TEMP must describe the kind and amount of suitability test and evaluation, required resources, and planned test locations and schedules. It must clearly relate T&E activity to suitability- related critical technical characteristics and operational suitability issues. It must describe the evaluation of the system relative to the suitability-related issues and the testing to be conducted to provide data to accomplish the evaluation. It must show the relationship between T&E schedules and program decision points and address the T&E to be accomplished in each program phase. It should identify the planned test articles to satisfy test objectives, as well as identify the number and rate of systems to be produced during the Low Rate Initial Production (LRIP) phase. Test resource requirements must be addressed, including known test resource shortfalls that may impede the full test and evaluation of the system suitability.

Finally, in reviewing a TEMP, one must be careful to avoid the pitfall of allowing the document to become an end unto itself. The document should define a test program that, when properly executed, will provide for accurate and efficient determination of a weapon system's operational effectiveness and suitability.

4.1 PART I, SYSTEM DETAILS

Part I of a TEMP provides details of the weapon system, its intended mission, and the required technical and operational characteristics. The mission must be adequately defined and key hardware and software features of the system must be described. Technical and operational characteristics discussions must include the relationship of those characteristics to the suitability aspects of system performance, as well as the effectiveness aspects.

AREA OF RISK

Heavy emphasis on effectiveness requirements may lead to oversights in suitability requirements.

The suitability implications of specified mission definitions and the system's technical and operational characteristics often are not well understood or clearly visible. During the early stages of a program, there is a tendency to place heavy emphasis on the definition and understanding of operational effectiveness. Later in the program, suitability deficiencies that significantly detract from system capability are identified.

OUTLINE FOR REDUCING RISK

The system's operational, support, and maintenance concepts should be examined to identify important suitability considerations, as well as the testability of key characteristics and requirements.

a.   **Mission Description. (Sec. 4.1.1)**

Can key operational suitability issues be identified from the mission description in the TEMP?

The mission of the system and the planned support concept should be described in enough detail to permit the reviewers and the decisionmakers to understand the critical operational issues (COIs), including the suitability COIs. If the mission contains new or unique requirements, these should be examined to ensure that the test program has the structure to deal with these features.

> The mission of a weapon system is to penetrate enemy defenses and to conduct offensive actions behind enemy lines in an independent and self-sustaining mode for a period of 60 days. Critical operational issues must address the suitability areas of reliability, maintainability, and graceful degradation of the system as required by the self-sustaining nature of the system's mission.

b.   System Description. (Sec. 4.1.2)

Are the key system features that drive the suitability requirements included?

Critical operational suitability-related system characteristics may be omitted, resulting in inadequate planning for that portion of OT&E. The suitability issues may be improperly identified; as a result, a critical issue will be missed. Inadequate attention might be placed on the operational suitability issues.

> A key feature of the missile warning system (MWS) is improved operational availability. With the use of fault-tolerant computer hardware and software, the system R&M will significantly improve end-to-end availability for the MWS. The improved R&M is critical to meet the high level of operational availability.

c.   Critical Technical Characteristics. (Sec. 4.1.3)

Do the technical characteristics support the operational suitability requirements?

The key technical characteristics should be described clearly. The rationale for each identified critical technical characteristic is important in understanding how the technical testing program fits into the overall acquisition program. This knowledge should then be used to understand the relationship of the technical testing to the key operational suitability requirements.

> The system has four critical areas of technical performance. The classified appendix lists these characteristics, and quantitative measures against test locations, schedule and the decisions supported, which provide the technical basis against which the system performance will be evaluated. These four areas are undetected message error rate, operational availability, processing time and system growth.

d.   Required Operational Characteristics. (Sec. 4.1.4)

Are the required operational characteristics and their associated parameters listed?

The operational characteristics--with associated operational effectiveness and suitability parameters--that are critical to the mission performance and the ability to place the system into field use should be listed. Thresholds, which represent the level of system performance acceptable to the user to successfully execute the mission, also should be listed. The Service-prepared TEMP identifies the key required operational characteristics needed for operational mission accomplishment. One must ensure that consideration is given to the testability of the requirements, utilizing the expertise of the development and test and evaluation communities, and check that the key required operational suitability characteristics deemed of critical importance to meeting the mission requirements are identified.

> The key required operational characteristics of the system include the following suitability requirements: system reliability, system maintainability, and logistics supportability. The values are contained in the classified appendix.

4.1.1 MISSION DESCRIPTION

The Mission Description section of the TEMP should briefly describe the mission of the deployed system, the threat it is required to be effective against, any threats that must be countered during the accomplishment of the mission, and the range of environmental conditions (weather, terrain, oceanographic, space) over which the system should be effective and suitable. As necessary, it should reference other appropriate, approved program documents. This might include the Mission Need Statement (MNS), System Concept Paper (SCP), Decision Coordinating Paper (DCP), or Service need statements, such as the Required Operational Capability Statement (ROC) or Statement of Operational Need (SON).

AREA OF RISK

The test program may not be structured to evaluate the system for its actual operational mission.

If the mission profiles are not clearly defined and understood by both the user and the developer early in the acquisition cycle, not only may the analyses and design activities be jeopardized, but also the test program may be planned in a manner that is not fully relevant to its intended use. In the suitability area, this mission definition needs to identify any specific reliability, maintainability, logistics, or other suitability area that has a critical relationship to the successful accomplishment of the system's mission.

OUTLINE FOR REDUCING RISK

Reducing the risk associated with the mission description and the associated suitability issues requires that the system's operational and support concepts be examined. Those factors that are important suitability issues should be identified and documented in the test and evaluation master plan.

a.   **Mission Definition.**

Does the mission description include items that could be suitability issues?

The mission or missions that are planned for the system should be described in sufficient detail to ensure that critical operational suitability issues can be identified, and the context of these issues understood. The TEMP should be compared with other program documentation to ensure that it agrees with the previous mission descriptions. If the detail is inadequate, then the need for additional detail should be judged, based upon what information is available from other sources and if this information is in documentation that is controlled as to content.

> The GBU-15 is a modular, unpowered, air-to-surface guided munition designed for external carriage on F-4 and F-111 aircraft. It is designed for employment from low and high altitude and appropriate standoff ranges against high value targets. The addition of an infrared guidance section provides day/night capability as well as limited adverse weather capabilities. Suitability areas of particular importance include: maintenance concepts, environmental considerations, safety, and transportation (including handling).

b.   Environmental Conditions.

Is the range of environmental conditions, over which the system will be effective, identified?

The mission description should include a discussion of the range of environmental conditions over which the system must be effective. This should include weather, oceanographic, space, terrain, obscurants, vegetation, illumination, etc., as is appropriate for the particular system in question.

> The vehicle loaded to the "full combat load" gross weight should be able to ford water to the depth of four feet and climb a 30 degree incline. The low-light-level rangefinder should be able to support the scout mission under the ambient nighttime conditions that are representative of conditions in Europe and the Middle East.

c.   Definition of Logistic Support Concept.

Is the logistic support concept defined in sufficient detail to allow the planning of the operational suitability portion of the OT&E?

The planned logistic support concept for the system may require that unique aspects or approaches for the support of the system be realized. The mission description in the TEMP should highlight any of these aspects that are unusual for the type of system being discussed. The need here is to identify any unique aspects of the support scenario or concept in enough detail so that the test program can be planned to address these aspects. The detail must be at a level that shows the unique aspects and gives the planners a foundation to structure the OT&E.

> The F-XX is a light attack aircraft with leap-ahead combat effectiveness and battlefield survivability to defeat the threat of the mid-1990s. The F-XX has a worldwide operational capability and high sortie generation rates greater than 5 sorties per day in sustained operation. It will perform combat tasks in the close-in, deep, and rear battle environments. The support concept for the aircraft is a three-level concept. In contrast, major avionics items have the reliability and maintainability required to be supported using two levels of maintenance. (Avionics items to be supported by two levels of maintenance are listed in table V-1.)

4.1.2 SYSTEM DESCRIPTION

The System Description section of the TEMP briefly describes the system's design including key features and unique characteristics, interfaces with other systems, and unique support concepts. Key features and subsystems should include both hardware and software elements, as appropriate. Unique characteristics or unique support concepts should be identified if they result in the requirement for special test or analyses during test and evaluation. The relative maturity, integration, and modification requirements of any Non-Developmental Items (NDI) should be addressed. Interfaces with existing or planned systems that are required for mission accomplishment should be identified. Any interoperability with existing and/or planned systems of other DoD Components or allies should be identified.

AREA OF RISK

A poor system description may lead to improperly identified suitability issues or missed critical issues.

The significance of the System Description, relative to operational suitability, is that critical system characteristics may be omitted and the requirements for the operational suitability portion of OT&E may be improperly planned. The suitability issues may be improperly identified; as a result, a critical issue will be missed. Test planners and the decisionmakers might place inadequate attention on the operational suitability issues.

OUTLINE FOR REDUCING RISK

It is necessary to review the supporting documentation and ensure that the system description adequately describes the system's operational suitability features, and what is different about this system.

a. **Key Features Description.**

Does the description of key features of the system and its subsystems include those that relate to the operational suitability issues?

Key features and subsystems are those that allow the system to perform its required operational mission. The descriptions of the key features and subsystems provide a basis of information for assessing the adequacy of the OT&E program that is described in the TEMP.

> The mortar consists of a 55mm tube, a telescopic sight, and a two-piece baseplate. Eight are issued to 20-man mortar platoons. The mortar is designed to be hand-carried by two men. The ammunition weighs 25 lbs. per round. The platoon will have sufficient HMMWV vehicles to transport every mortar plus its basic load of ammunition (50 rounds).

b.   Relationship with Existing or Planned System(s).

Are the interfaces with existing systems identified?

The identification of the interfaces with other existing or planned systems results in a list of potential requirements for interoperability and/or compatibility issues. This list forms the basis for systems that must be acquired for use during the operational testing. If the systems are not available when required, then the potential for a serious test limitation must be examined. The electromagnetic environmental effects (E3) area can be a significant issue for some systems. Having complete knowledge of all other systems that will be operating in proximity to the system under test will help in defining the test objectives.

> Two communications systems, each with a different mission, entered development at approximately the same time. The first was subjected to operational testing without the second system being present. When a joint test finally was conducted, there was a serious incompatibility problem between the two systems. It was not possible to operate the first system unless the second system was turned off.

c.   Unique Characteristics of the System.

Are unique characteristics of the system or unique support concepts identified?

Characteristics that are unique, different, or better in relation to operational suitability should be identified. Any unique suitability characteristics should be considered when the suitability COIs are defined. To form a basis for this consideration, these unique characteristics should be listed in the System Description. Also, these unique characteristics or support concepts should be considered when planning for the operational test events. Having an adequate description in the TEMP allows the remaining planning to proceed.

> A missile system has a unique Go/No-Go test to indicate the status of the system. It is important that the test be designed to clearly isolate a fault to the missile or the launching system. Failure to isolate the fault in this manner will result in delayed or ineffective missions or the need for additional maintenance capability. The TEMP describes these features in sufficient detail to ensure such unique tests are addressed in the test plan.

d.   Changes over Previous System.

Are there significant changes or improvements over the predecessor system?

Major improvements in performance over the predecessor system can result in risk areas that should be the focus of operational testing. If such improvements are predicted in some of the suitability elements, the remaining elements will be adjusted to remain in balance. Thus, the major improvement must be confirmed during OT if the system is to be operationally suitable.

> An electronics system using new technology electronic components is forecasted to have a Mean Time Between Failure that is seven times that of the system it is replacing. The maintenance manpower and the logistics support that is planned are significantly less than that of the predecessor system. The reliability of the system is identified as a COI.

4.1.3 CRITICAL TECHNICAL CHARACTERISTICS

The Critical Technical Characteristics section identifies the technical characteristics whose measurements are the principal indicators of the system's technical achievement. The section should identify performance thresholds and the milestones at which each of the thresholds has been, or is scheduled to be, demonstrated. Characteristics should be quantified when possible. Other program documents may be referenced for the technical characteristics, particularly if the technical requirements are classified. Technical characteristics usually are contract specifications and are derived from operational user requirements which precipitated the need for the system. Therefore, they may be traceable to the required operational effectiveness and suitability characteristics.

AREA OF RISK

Poor definition of critical technical characteristics can lead to a test program that is unable to assess program risk.

TEMPs frequently do not demonstrate the relationship between the technical characteristics and the operational characteristics, including operational suitability. The technical thresholds provided oftentimes are presented without a clear justification for their choice or an explanation of their significance or testability prior to the intended milestone. The purpose of technical testing is to reduce risk by assuring that a portion of the challenge of achieving the operational requirements has been met. To understand what degree of risk has been reduced, the relationship between the technical characteristics and the operational requirements must be discussed.

OUTLINE FOR REDUCING RISK

Having an understanding of the technical characteristics and the rationale or specific reference relating each identified critical technical characteristic is important to understanding how the technical testing program fits into the overall acquisition program. This knowledge should then be used to understand the relationship to the required operational suitability capability.

a.   **Critical Characteristics Listing.**

Are the key technical hardware and software characteristics and thresholds listed?

Key technical characteristics are found in the system specifications and requirement documents. The TEMP should summarize them and relate them to the milestones in a matrix format. The matrix should indicate the characteristics that have been evaluated or that will be evaluated during the remaining phases of the developmental test (DT). Each technical characteristic should have a threshold value. If the technical characteristics are summarized in another acquisition document, such as a Baseline Correlation Matrix (BCM), this document should be compared to the TEMP. The technical characteristics that relate to operational suitability generally are quantifiable items, such as reliability and maintainability. One should examine these characteristics and determine how they relate to the operational suitability characteristics. What is their basis? Are they engineering estimates of expected performance required by contract specifications?

Since technical characteristics often are measured in a more pristine manner than exists in the operational units, the reliability and maintainability characteristics tend to be more optimistic than the operational characteristics. They should be examined to determine if the proper relationship exists.

CRITICAL TECHNICAL CHARACTERISTIC [Measurable Characteristic with Reference]	TEST EVENTS [Single Event or Test Phase]	TECHNICAL THRESHOLD FOR EACH TEST EVENT [Measurable Technical Achievement]	LOCATION [Test Facility]	SCHEDULE [Test Period]	DECISION SUPPORTED [Milestone, In-Process Review, or Major Event]
Reliability - Mean Time Between Operational Mission Failures (MTBOMF)	RGT I	200 Hours	Contractor's	1/92 - 7/92	
	RGT II	325 Hours	Contractor's	10/92 - 3/93	Service IPR
	FSD SYS DT	455 Hours	Test Bed	7/93 - 11/93	MS IIIA

b.   **Milestone Intervals.**

Are milestones for demonstration of key suitability-related technical thresholds identified?

The technical thresholds should be tied to specific milestones. At those milestones, sufficient testing of the respective technical areas should have been accomplished and the results submitted to the decisionmakers. This schedule also should be compatible with the operational T&E of the companion operational suitability characteristics. The program milestone documentation should be checked to ensure that the technical characteristics that are important to critical suitability issues are addressed on a schedule consistent with the rest of the acquisition program.

> The Technical Characteristics listing indicates the decision that is supported by each of the technical thresholds. Phases of testing are indicated that will yield information to judge the system against the thresholds at each decision point.

c.   **Suitability Characteristics.**

Do the technical suitability characteristics describe minimum acceptable system performance?

The technical characteristics should be quantifiable, where appropriate, related to system operational suitability characteristics, and provide the quantitative measures against which system performance will be evaluated during system-level technical testing. Technical suitability characteristics usually are defined differently than operational characteristics, and, as such, result in higher values.

> The Technical Characteristics listing indicates that the diagnostics false alarm rate threshold is 2 percent of all system test failures. For two BIT actuations, the false alarm rate shall be less that 0.1 percent. Diagnostics between different levels of testing shall incorporate tolerance levels that preclude retest-OK and cannot duplicate anomalies above the 2 percent rate.

4.1.4 REQUIRED OPERATIONAL CHARACTERISTICS

This section of the TEMP should list the key operational characteristics and their associated parameters as identified in the user's need statement and approved by the Service. A key operational characteristic is a principal element of the system's ability to accomplish its mission (operational effectiveness) and to be supported (operational suitability). These characteristics usually are defined by parameters that are indicators of the system's ability to accomplish its mission. If the Service has an approved matrix of operational requirements in the requirements documents, it may be used to display the required operational characteristics.

AREA OF RISK

Key operational characteristics may not be highlighted and the planned testing may not be sufficient to ensure achievement of the mission requirement.

Key required operational characteristics needed for operational mission accomplishment may address attention to the deficiencies of the current system. While characteristics often are drawn from Service-approved formal documents, the danger exists that the formal document may be incomplete or may be written in generalized, nonquantitative terms. Consequently, the operational requirements parameters selected for highlighting may not be sufficient to ensure achievement of the mission requirements. Since the operational suitability characteristics are seen by many people as secondary considerations, there needs to be assurance that adequate description and definition of the these items are included in the TEMP.

OUTLINE FOR REDUCING RISK

Key required suitability characteristics that are of critical importance to meeting the mission requirements should be identified. The description of each key suitability characteristic should include the appropriate parameters and thresholds.

a.   Operational Suitability Requirements.

Are all operational suitability requirements identified?

Suitability issues are identified in the TEMP. The COIs should cover all suitability elements; the reason for eliminating any element should be discussed. Other operational requirements might be in the diagnostics area or in levels of support needs, e.g., numbers of maintenance people, test equipment, levels of maintenance, and software change requirements. In reviewing the TEMP, the key is to ensure that the critical suitability areas are defined by operational requirements. Definitions should include all operationally relevant situations; e.g., all failures that can occur in operational use should be included in the definition for the operational suitability characteristics and parameters. This information should be a direct follow-on from the system's user requirements. It is very difficult for the operational test community to plan and conduct a meaningful operational test if the user community has not completely defined realistic quantitative needs.

> The reliability of the communications system should be indicated by the mean time between operational mission failure (MTBOMF) values demonstrated. The user equipment-requirement MTBOMF must be greater than 500 hours for both the manpack and the aviation sets. The test is planned to demonstrate an 80 percent confidence in the MTBOMF values.

b.   Parameters and Thresholds.

Are suitability characteristics supported by parameters and thresholds?

Each of the key operational characteristics should have an associated parameter, and the parameters should be accompanied by a threshold (usually but not always quantitative). The thresholds should not be engineering estimates of expected performance or the performance that is required by contract specifications; they should be the minimum system performance acceptable to the user to successfully execute the mission. If only a component of a system is being developed, the thresholds should represent the essential operational requirements of the total weapon system. Failure definitions should be included by reference, and should reflect all failure modes or events that can be expected to occur in operational service.

REQUIRED OPERATIONAL CHARACTERISTICS WITH THEIR ASSOCIATED PARAMETERS AND THRESHOLDS		
CHARACTERISTIC	PARAMETER	THRESHOLD
Reliability	Mean time between operational mission failures (MTBOMF)	500 hrs.
	Mean time between maintenance actions (MTBMA)	32 hrs.
Maintainability	Maintenance manhours per operating hour	2.5 hrs.

c.   Qualitative Suitability Requirements.

Are critical suitability areas expressed qualitatively?

With some systems, there are critical aspects of the system's suitability performance that can be expressed qualitatively only. There should be no reluctance to include such items in the list of required operational characteristics. The items that are critical to the system's ability to perform its mission within the intended field environment may require objectives that are describable only in a qualitative manner. These could include requirements for compatibility with other systems or with certain skill level operating and maintenance personnel, or requirements that are in areas that have not progressed to the level where the requirement can be accurately quantified.

> The system must be air transportable. This requires that the system must be able to be prepared for air shipment by organizational level personnel and the system be air transportable by C-130, C-141, and C-5 aircraft.

4.2 PART II, PROGRAM SUMMARY

Part II of a TEMP identifies responsibilities of the participating organizations, as well as the management and scheduling aspects of the weapon system program. Key factors to be discussed are T&E and acquisition strategy relationships, key decision points and associated reports, T&E requirements to support LRIP, constraints, test article and critical support resource availability, and the associated T&E responsibilities of all participating organizations.

AREA OF RISK

Suitability testing will not be performed at a time to support the program's decision milestones.

There is significant program risk if the suitability of the system is not demonstrated in time to support decision milestones. Typically, the operational suitability aspects of a weapon system program lag the effectiveness aspects of the program. Design of support equipment cannot commence until the design of the actual weapon system begins to stabilize. Maintenance manuals cannot be written until the support equipment is defined. As a result of this inherent lag, actual testing of the suitability aspects more often than not does not occur in concert with testing the effectiveness of the weapons system. If not appropriately planned for and scheduled, critical suitability elements, including diagnostics capabilities, support equipment, technical manuals, etc., may not be available. Therefore, suitability testing will not be performed in a timely manner to support the program's decision milestones.

OUTLINE FOR REDUCING RISK

Management responsibilities must be clearly delineated and include emphasis on the suitability aspects of the planned test program. The integrated schedule should indicate that the test program is not schedule driven, but accomplishment driven.

a.　**Integrated Schedule. (see 4.2.1)**

Are all the required suitability activities clearly identified, appropriately time-phased, and adequately resourced so as to provide the required results at the decision milestones?

The integrated schedule should demonstrate that suitability testing will be consistent with the need for information at the key program decision milestones. The relationship of the test periods to the schedule should give an idea of the amount of test time that will be available to support the various milestone decisions and the associated reporting requirements. The integrated schedule should give an indication of when the various suitability elements will be available for OT&E; e.g. when the support equipment and representative maintenance personnel will become available for the various levels of support, etc. The placement of these items on the integrated schedule will give the reviewer an idea of when various portions of the operational suitability evaluation will be done and when the results of this activity will become available for presentation to the decisionmakers. The schedule must be reviewed to ensure the needed support to obtain suitability objectives has been identified; this includes testing and associated hardware and facility requirements down to the lowest practical level.

The integrated schedule for the M165 Truck program displays "on-dock" dates for the test articles to be used for maintenance tear-down and test personnel training. The sequencing of the events indicates that the required support elements will be available in a timely manner to support operational testing.

b.   **Management. (see 4.2.2)**

Are responsibilities for the operational suitability areas clearly identified and appropriately assigned?

One must ensure that the responsibilities are depicted, giving suitability objectives the needed attention, and that a sufficient number of test events are scheduled to provide a level of confidence in the resulting suitability statistics. Also, the relationship between the key decision points and the specific T&E reports should be examined.

Overall responsibility of the LX-21 Helicopter program is the responsibility of the Program Manager (PM). The PM establishes and chairs the TIWG and assures that adequate technical testing is accomplished. TRADOC represents the user and also is responsible for ensuring that the training of all test players meets or exceeds the minimum acceptable standards used to measure training effectiveness. Technical testing under TECOM's direction, includes...Reliability, Availability and Maintainability (RAM) measures. AVSCOM also is responsible for management oversight...to assure...RAM maturity and growth is continuing as it pertains to flight handling qualities and airworthiness. U.S. Army OTEA will conduct Continuous Comprehensive Evaluation and the initial OT&E using LRIP aircraft. USALEA is responsible for...ensuring appropriate logistics testing and evaluation are accomplished.

4.2.1 INTEGRATED SCHEDULE

The entire program schedule should be displayed on one page to include the integrated time sequencing of critical T&E phases or events, related activities, and planned cumulative funding expenditures by appropriation. Included should be event dates such as program decision milestones, test article availability, critical support resource availability; appropriate phases of DT&E, live fire T&E, and OT&E; rate production deliveries (i.e., low and full); Initial Operational Capability (IOC); Full Operational Capability (FOC); Low-Rate Initial Production (LRIP); and beyond LRIP.

AREA OF RISK

Insufficient time and resources may be planned for suitability testing.

The integrated schedule may not cover all the necessary events or dates to communicate all of the aspects of system testing. The TEMP integrated schedule may be at such a summary level that it is difficult to determine the time and resources available for the suitability portion of the OT&E. Without this level of visibility, it might not be possible to assess the planned testing for the suitability objectives. In addition, critical suitability assets may not be available when required during the testing period. Such factors could result in the non-completion of testing and increase the risk of making inappropriate decisions at the decision milestones.

OUTLINE FOR REDUCING RISK

The compatibility of the schedule with the program decision milestones and the compatibility of the hardware development and testing with the software development and testing also should be evaluated. One should review the supplementary schedules and data source matrix (DSM) in the TEMP, as well as the timing sequences relating software test to hardware testing. The integrated schedule should be examined to determine if the time and resources that are available for the suitability portion of the OT&E are adequate.

a.   **Schedule Supports the Program Milestones.**

Has sufficient time been scheduled to ensure the collection of meaningful suitability-related data? Will results be available prior to the requisite milestone?

The integrated schedule should demonstrate that the testing will yield suitability results at a time consistent with the need for information at the key program decision milestones. The relationship of the test periods to the program schedule should give an idea of the amount of test time that will be available to support the various milestone decisions and the associated reporting requirements. (The details for suitability testing are discussed in sections 4.4 and 4.4.4.)

The system IOT&E consists of 750 flight hours and will be completed at least 120 days prior to Milestone III; thus, there should be many opportunities for maintenance actions. The test phase will provide adequate maintenance data to present the suitability results to the Milestone III DAB.

---

b.   Scheduling Suitability Elements.

Does the schedule include an on-dock date for support equipment?   Will maintenance personnel training be completed in time to support OT&E maintenance-related testing?

The integrated schedule should give an indication of when the various suitability elements will be available for OT&E; this includes when the support equipment will become available for the various levels of support, when representative maintenance personnel will become available for the various levels of support, etc. The placement of these items on the integrated schedule will give the reviewer an idea of when various portions of the operational suitability evaluation will be completed and when the results of this activity will become available for presentation to the decisionmakers.

> For the IOT&E, the first- and second-level maintenance of the system will be performed by representative military personnel. Prototype test and support equipment will be available for the test phase. This phase will provide the first operational testing of the test equipment.

c.   Scheduling Adequate Time.

Is adequate time scheduled for the suitability portions of operational testing?

One should review the schedule to ensure that the support elements required to meet the suitability objectives have been identified; this includes testing and associated hardware and facility requirements down to the lowest practical level. If program schedules have slipped, then rescheduling of reviews should be examined to ensure that adequate time for OT&E conduct is planned prior to the scheduled review.

> The 750 flight hours that are planned for IOT&E will provide a sufficient number of the two designated system missions to demonstrate the system's reliability for the missions. Scheduled "on-dock" dates for support equipment, spares, T.O.s, and training are supportive of the planned start of flight testing.

d.   Software.

Are adequate time and necessary resources scheduled for the planned operational testing to allow for software testing concurrent with hardware testing?   Will the software be baselined and under configuration control prior to the start of OT&E?

One must ensure that the TEMP includes key software releases, subsystem/system tests, and sufficient key system events to coordinate software testing with the system schedule. The calender time available for software testing should be indicated. Resources necessary to support software testing must be planned for and scheduled.

> The integrated schedule provides visibility of the schedule for planned software block changes and clearly indicates a logical time-phased sequence for their introduction into the OT&E process in a controlled and organized fashion.

4.2.2 MANAGEMENT

The purpose of the Management section is to outline the T&E responsibilities of primary participating organizations (developers, testers, evaluators, and users). The T&E strategy should be related to the acquisition strategy of the program (any concurrency of production and testing should be discussed). The key decision points should be listed, along with the T&E reports that will support those decisions. Terms such as "Low Rate Initial Production," "Full Rate Production," and "Initial Operational Capability" should be defined quantitatively (rate and total quantity). The scheduled date (e.g., fiscal year quarter) for the decision to proceed beyond LRIP should be identified. The management of schedule, resource, or budgeting constraints that may impact the adequacy of planned testing should be addressed.

AREA OF RISK

Inadequate emphasis may be placed on addressing the suitability issues of the OT&E.

The risk to operational suitability in the T&E management areas is that the management structure of the program may place inadequate emphasis on the suitability issues of the T&E. Some managers tend to focus attention on items that are important to evaluating operational effectiveness and pay little attention to the operational suitability objectives. This management emphasis is the result of the chronological sequence that places effectiveness before suitability.

OUTLINE FOR REDUCING RISK

Reduction of risk requires that the manager be held accountable for both effectiveness and suitability objectives over the life of the test. Accordingly, these responsibilities are to be clearly identified in the TEMP. Resources and schedules should provide the capability to perform the required testing.

a.   Management Responsibilities.

Is there an outline of management responsibilities to ensure suitability objectives receive the proper attention?

The responsibilities that are spelled out should give the suitability objectives the needed attention. The proposed management team and approach will indicate the organization that will be conducting and contributing to the T&E in the suitability area. In some cases, suitability elements may be evaluated by supporting organizations. These organizations need to be identified in early versions of the TEMP to ensure that resources are available when required.

> The Test Integration Working Group (TIWG) will interface with the following functional groups: the Integrated Logistics Support Management Team (ILSMT), the Training Support Working Group, the MANPRINT Joint Working Group, the Computer Resources Working Group, and the Operational Test Readiness Review Working Group.

b.   Proposed Testing.

Is it clear that testing is adequate in scope to provide confidence in the planned suitability results?

A sufficient number of test events should be scheduled to provide a level of confidence in the resulting suitability statistics. If significant levels of risk are the result of limited test assets, then these risks should be discussed.

> The IOT&E, as currently estimated, consists of ground tests, jettison/separation tests, captive flights and the launch/flight testing of 17 missiles using the F-18 as the carrier aircraft.

c.   Schedule Compatibility.

Is the proposed relationship between the decision milestones and the T&E reports compatible?

The management section of the TEMP should indicate the planned sequencing of T&E reports from testing phases that support the major program milestones. The definitions of the program milestones will provide a context for assessing the adequacy of the suitability information that will be available at that milestone decision point.

> The schedule for an aircraft test and evaluation indicated that the test articles would be delivered to the Service DT&E on 1 June, where DT would be conducted until 1 August. The test articles were to be delivered to the OT&E site (400 miles away) on 4 August for 90 days of IT&E. The Milestone III was scheduled for 10 December. This schedule did not allow sufficient time for the deficiencies found in DT to be corrected before OT and did not allow sufficient time to complete the evaluation at the end of OT&E before the milestone decision.

4.3 PART III, DT&E

Part III of a TEMP is devoted to the development testing and Evaluation of a weapon system program. Responsibility for its content rests with Deputy Director, Defense Research and Engineering (T&E). This section should summarize those activities planned for the development test phase(s). Knowledge of those activities can be of value to the DOT&E staff assistant in understanding the overall testing concepts and potential availability of data to accomplish early operational assessments.

4.4 PART IV, OT&E

Part IV of the TEMP focuses on the Operational Test and Evaluation portion of the overall test program. It highlights the Critical Operational Issues (COIs), summarizes the OT&E that has been performed to date, and describes the OT&E that is planned.

AREA OF RISK

An ineffective test program may be structured.

The OT&E description may lack detailed information or communicate information that is unclear and ambiguous. Critical Operational Issues (COIs) may not be identified. Problems or limitations with past OT&E may not be included in the OT&E summary. Limited definition of the planned OT&E may result in the T&E "contract" being assumed to be one thing, while the plan is for something else.

OUTLINE FOR REDUCING RISK

a.   **OT&E Overview. (see 4.4.1)**

Is the OT&E overview complete?

This section should provide a summary of how the OT&E is structured. It should show how the program structure, test management structure, and the required resources are related to the system requirements, COIs, test objectives and decision milestones. It also should show how the completed OT&E has evaluated the system and how the future OT&E will evaluate the system. This overview should give adequate attention to the suitability COIs and discuss how the suitability evaluations/assessments will be provided for each of the decision milestones.

> A review of the TEMP indicated that the OT&E overview was an attempt by the system developer to justify performing a scaled-down OT&E.

b.   **Critical Operational Issues. (see 4.4.2)**

Are all COIs identified?

Factors which could preclude suitability performance, as required by the users, must be identified. Identification and description of the related COIs are required to focus test resources and the attention of the decisionmakers on these important issues. Program aspects that may result in suitability-related COIs include a statement of higher levels of suitability performance than that required in previous similar systems, as well as the introduction of new or unproven technology into a system.

> Can the units that are equipped with the system achieve their peacetime and wartime system readiness objectives (SROs)? The system shall demonstrate a peacetime operational availability of 0.86 and a wartime operational availability of 0.78. The probability of successfully completing a three-hour mission shall be 0.70, with a mean time between operational mission failure of 8.5 hours.

c.   **OT&E to Date. (see 4.4.3)**

Is the summary of OT&E to date complete and accurate?

The description of prior OT&E should not be a duplication of the detailed OT&E reports. The discussion should summarize the prior operational tests, including what portions of the support system were tested and what the results were; what suitability COIs have been fully or partially addressed; what the results were; what planned activities were not performed, and why.

> The IOT&E was initiated in January of 1985 and was conducted for a two-month period. Insufficient data were accumulated to provide quantitative measures for reliability or maintainability. Five qualitative maintainability deficiencies were documented in the report for this short test phase.

d.   **Future OT&E. (see 4.4.4)**

Does the discussion of future OT&E include a complete description of each OT&E test phase?

All future OT&E should be described and its specific purpose identified. Any major deficiencies should be addressed, as well as when their correction will be verified. Each phase of OT&E should be discussed separately. The configuration to be tested in each phase of OT&E should be identified. The portion of the support system that is present in each of these test phases should be discussed. The suitability objectives for each phase of OT&E should be listed. A brief description should portray how each phase of testing will be conducted (events to be performed, types of representative support personnel to be used, how the system maintenance and logistics support will be evaluated in this phase, and the role of suitability modeling and simulation in this phase). Those factors that limit the full and completely realistic operational test of the suitability aspects of the system should be identified.

> The IOT&E supports Milestone III and will consist of a 60-day, 300-flying-hour effort using three prototype aircraft. The test will be conducted between July and October 1990. Logistics and maintainability demonstrations will be conducted to provide an early assessment of the user's ability to maintain the system under a concept of two maintenance levels.

4.4.1 OT&E OVERVIEW

The OT&E Overview should provide a summary of how the OT&E is structured. It should show how the program structure, test management structure, and the required resources are related to the system requirements, suitability COIs, test objectives and decision milestones. It should show also how the completed OT&E has evaluated the system suitability and how the future OT&E will evaluate the system. This overview should give adequate attention to the suitability COIs and discuss how the suitability evaluations/assessments will be provided for use in each of the decision milestones.

AREA OF RISK

An inadequate description leads to the approval of an unacceptable suitability test and evaluation program.

If the discussion in the OT&E outline is inadequate, or if the coverage of operational suitability is not adequate, then the information that is used to judge the acceptability of the TEMP and the test program also will be deficient. As a result, the test program that is approved may be unable to meet the needs of the decisionmakers.

OUTLINE FOR REDUCING RISK

To reduce the risk associated with inadequate testing, it is important to have an understanding of the testing already conducted and that which is planned. The OT&E Overview should include the contractor testing and early technical testing, as well as early operational test and evaluation. The operational evaluation must take advantage of testing data from all appropriate sources up to, and including, the independent operational test and evaluation. Follow-on testing should be used as soon as it is performed to assist in the evaluation of the production articles. The Overview indicates the level of development of test articles being used in the operational test; for example, LRIP or prototypes. If the test articles will not be full-rate production articles, there should be a discussion of the differences between the test and full-rate production articles and the effects of their use during OT&E.

a.   **OT&E Overview.**

Is the OT&E Overview in the TEMP a complete summary of the suitability T&E program?

The OT&E Overview should provide a clear understanding of the testing already completed, the testing yet to be completed, and who will process and evaluate the data. The overview should be organized by acquisition phase and specific information requirements; for example, information to the system's developer should be annotated. Special suitability testing requirements should be discussed; for example, the impact on the test of an inadequate test support package, an independent contractor to perform independent verification and validation (IV&V) on the system's software, or the use of the developing contractor to operate or maintain the tested system during the IOT&E.

The pre-Milestone II TEMP for an aircraft indicated in the OT&E Overview that an effort would be made to conduct early operational capability tests to provide the user with a perspective of the potential effectiveness and suitability during system's development. It was stated that OT evaluators will participate in contractor and technical testing exercises, to include demos, surveys, and mock-ups. The Overview stated that an independent contractor would be used to conduct IV&V, and this contractor would be available to the OTA to assess systems' software suitability. Because operating the system in a chemical environment is a major concern, the Overview provided a discussion of the operational testing of system's maintenance while in this environment. The Overview also stated that the IOT&E would be conducted using LRIP aircraft (with a brief description of the difference between LRIP and Full-Rate production systems), and that resulting data would support the full-rate production decision at Milestone III.

b.   OT&E Summary.

Does the DOT&E Overview summarize the OT&E that has already been conducted, to include test articles descriptions and the future OT&E?

The Overview should discuss the OT&E conducted to date, by phase, and outline the successes and failures in achieving the operational suitability characteristics. If reviews and decisions were made that altered future system development or affected testing, there should be a discussion of how these issues will be addressed. The Overview should provide a test article description that states the level of development of the tested system, and a synopsis of the OT&E events, with a discussion of the results of each event. Finally, the Overview should provide a summation of program management decisions that will impact on the acquisition schedule or operational testing and thereby require adjustments in test resources or test design and execution.

The future OT&E planned for the system must be applicable for the phases of the system's acquisition and in the time frames required. This section should provide a test article description of the system that will be tested in each of the future phases and a discussion of the OT&E objectives for each test. There also should be a detailed discussion of the OT&E events and scope of the testing, as well as a discussion of the basic scenarios that will be followed for each of the tests.

In evaluating the TEMP for a system, it was discovered that the future OT&E section stressed the completeness of past testing and attempted to justify a lack of planned future suitability OT&E for the system. The TEMP stated that because Service personnel would be participating in the DEM/VAL phase, have access to contractor IV&V data, and participate in demos, IOT&E should be reduced in scope to a 30-day field exercise. The TEMP proposed concurrent DT/OT prior to Milestone III, with each test organization having equal access to results. Finally, it proposed that an Early Operational Capability Unit participate and maintain the system in IOT&E because this would provide a head-start on the training of the Initial operational capability unit. Many of these statements tend to dilute the operational test by reducing the test time and resources, and by providing non-typical user troops.

4.4.2 CRITICAL OPERATIONAL ISSUES (COIs)

The Critical Operational Issues (COIs) are key operational effectiveness and operational suitability issues that must be examined in OT&E to determine the system's capability to perform its mission. The COIs are not characteristics, parameters, or thresholds, but they may have associated characteristics, parameters, or thresholds. COIs should cover all areas that critically affect the system's ability to accomplish its mission in the intended environment. The TEMP should identify which phase of OT&E will address each COI.

AREA OF RISK

Adequate attention may not be focused on some specific point that is important to the successful fielding of the system.

The significance of the COIs is that they help focus resources and management attention on items that are important in evaluating the system's progress toward attaining its operational objectives. The COIs allow the decisionmakers to have a smaller set of issues to address when making decisions on the acceptability of system development to date. Within this context, the risks are that the COIs may be improperly identified, or that a critical issue will be missed and the decisionmakers will not focus adequate attention on some point that is important to the successful fielding of the system.

OUTLINE FOR REDUCING RISK

The Critical Operational Issues (COIs) are the critical aspects of the system's operational effectiveness and operational suitability that are intended for examination and resolution during OT&E. Critical operational issues are developed by the tester and may be represented as questions that must be answered at the next acquisition decision milestone. The COIs are not characteristics, parameters, or thresholds, but they may have associated characteristics, parameters, or thresholds. The issues should cover all areas that critically affect the system's ability to accomplish its mission in the intended environment. The TEMP should identify which phase of OT&E will address each COI.

The emphasis of COIs is on the determination of the attainment of certain key performance levels and on surfacing potential problems that could interfere with successful mission accomplishment. Critical operational issues may change from one milestone to the next as some are resolved and new ones emerge in keeping with the systems development status. COIs must be structured to ensure that the information needs of the acquisition review body can be addressed for the milestone at hand.

Reducing the risk associated with OT&E COIs requires that the system's requirements, mission, and operating and support concepts be understood. Factors that are important critical issues related to operational suitability should be identified and documented in the TEMP. The full range of the intended operational environment must be considered. The list of COIs should be thorough enough to ensure that, if every COI is resolved favorably, the system should be operationally suitable when employed in its intended environment by typical users.

a.   Completeness of List of COIs.

Is the list of suitability COIs complete?

The major risk in this area is the situation in which some important suitability area has been overlooked and thus not identified in the list of COIs. As a result, attention is not given to this area when the Operational test plan is prepared, and the operational testing provides inadequate data to evaluate the condition at the time of an important milestone decision. The review of the TEMP should focus on identifying any critical suitability area that is not included in the list of COIs. Suitability risk areas may be identified from the areas of highest risk, or from the areas that are of the highest criticality within the support plans for the system.

> A remotely piloted vehicle (RPV) test program was planned to address three primary objectives, which were labeled as "critical" issues. Other issues including survivability, RAM, training, and human factors were to be addressed only to the extent that they affected the RPV's ability to meet the criteria of the three "critical" issues. Program milestone documentation did not contain explicit suitability criteria. The approach to suitability testing was to observe the ability of the RPV to support the mission under sustained combat operations, noting any shortfalls that could be attributed to suitability problems. This approach was justified by arguing that only "critical" issues required explicit criteria, and, therefore, suitability did not require explicit threshold values.

b.   Suitability Requirements.

Are the suitability requirements at high levels compared to previous systems? Are they identified as COIs?

The planned levels of suitability performance (particularly reliability and maintainability) may be COIs if the system's successful operation is dependent on achieving a markedly higher level of reliability, maintainability, etc.

> The reliability requirement for the new system is significantly higher than the existing system (failure rate is one-third of the previous rate). A COI has been identified as "has the system achieved the planned level of reliability?"

c.   System Technology Risk Areas.

Is new or unproven technology required for this system, or planned for use? Are there COIs on these technologies?

The use of advanced technologies in a system may introduce risk both in achieving system performance levels and in achieving the capability to support such new technology. Are there relatively unproven technologies in the system? Are these technologies understood from a reliability standpoint? Is the process to support these technologies (maintenance procedures, test methods, maintenance training, human factors, etc.) understood or demonstrated?

> A new, unproven cooling method is proposed for the IR seeker of a guided missile. The reliability of the cooling and the seeker is critical to the missile's operation. Therefore, the reliability and maintainability of the seeker and its cooling are specified as a COI.

4.4.3 OT&E TO DATE

Each completed phase of OT&E should be summarized. Descriptions of the hardware and software actually tested should be provided. Differences between the system used in testing and the configuration expected to be fielded should be highlighted, and potential impacts to suitability resulting from the differences should be discussed. The actual suitability testing that occurred should be summarized, including events, scenarios, resources used, test limitations, results achieved, and the evaluations conducted. Planned suitability objectives that were not met should be highlighted and explained. Status on the resolution of all suitability-related Critical Operational Issues should be discussed.

AREA OF RISK

The suitability-related requirements that remain for the future OT&E phases may be misjudged.

Discussions may concentrate more on the operational effectiveness requirements and ignore the suitability requirements. In some cases, the information may be condensed to such a degree that there is very little information provided to form a basis for judging the suitability-related requirements that remain for the future OT&E phases. If numerous phases of OT&E have been completed, the appropriate OT&E reports may be referenced and only the most recent or pertinent OT&E results included in the TEMP discussion.

OUTLINE FOR REDUCING RISK

Reducing the risk associated with suitability requires that the system's suitability-related requirements be tested. The test results should be examined to ensure the objectives of availability, compatibility, transportability, safety, human factors, interoperability, reliability, wartime usage rates, maintainability, manpower, training, supportability, logistics supportability, software supportability, and documentation have been met.

a. **Summary of Actual Testing.**

Is the previous testing for suitability summarized?

The previous testing should be summarized to include events, scenarios, resources used, test limitations, evaluations conducted, and results achieved. Specific OT&E reports that contain the detailed results should be identified. This summary should address the suitability areas adequately and should not be a summary of the effectiveness testing to date. Suitability elements that are included in each of the test phases should be identified.

> During phase IA of the OT, the suitability evaluation did not address the direct level of maintenance support, because the test equipment that was to be delivered to the test site was not available in time to meet the dates that were mandated by the scheduled range times.

b.   System Configuration.

Is the configuration of the test systems used in previous phases identified?   Were similarities and differences with production or later test articles summarized?

The early phases of the testing are necessarily performed on systems that are not of the production configuration. Nonetheless, the data from these test phases are valuable. The key is to place the test results into context with the configuration of the test articles and the realism of the test environment. The summary of the prior OT&E phases should document these factors.

> The systems used in the Phase I IOT&E did not have the diagnostics software included in operational software for the mission computer.   Therefore, the diagnostics objectives were not evaluated.

c.   Suitability Objectives.

Are the suitability objectives of the prior OT&E activity described?

The discussion of prior activity can be relatively brief or can include references to previous reports. The critical questions are: what suitability objectives were planned to be addressed by the prior test phases, and what were the results? Is there a discussion of significant events, test conditions, scenarios, resources used, limitations, and results as related to the system's suitability requirements?

> Phase I IOT&E was conducted on two of the initial prototype systems. Reliability data were collected during the test phase. Since the DT&E test team provided the system maintenance, no maintainability data were collected.   A qualitative assessment was made by observing the maintenance as it was performed.

d.   Suitability Results.

What were the results of the prior OT&E phases in the area of operational suitability?

The previous operational testing phases should be summarized and should provide insight into the suitability issues that have been addressed and the results that were achieved. A listing of the COIs (including the suitability COIs) should be included. This listing should indicate which of the COIs were resolved (satisfactory, unsatisfactory, yes, no, etc.), partially resolved, or unresolved at the completion of that phase of testing.

> The threshold for this time frame was established as a system reliability of greater than 0.80, so the issue was satisfactory for the Phase I OT&E. (Since the thresholds will change for the various phases of operational testing, this success does not mean that the suitability will automatically remain satisfactory.   The thresholds will become more demanding in the later phases of OT.) In the case of the maintainability COI, the qualitative maintainability assessment concluded that the access provisions for the required preventative maintenance tasks were unacceptable. Since the design of the access provisions were not projected to be revised in later development configurations, this deficiency was highlighted as a potential limitation on the system's operational suitability once it was fielded. The program manger was directed to resolve the deficiency.

4.4.4 FUTURE OT&E

All remaining OT&E required to resolve suitability-related Critical Operational Issues should be discussed. Operational testing should be described to verify the correction of major suitability deficiencies that were previously identified. A major deficiency is one that precludes the system from being designated as "operationally suitable."

AREA OF RISK

Testing may be structured so that a critical suitability issue will be missed.

Future testing may not be adequately described or the OT&E may not be directed at some key issue that must be examined before the appropriate decision milestone. The results of earlier testing may identify the need for testing of deficient areas that have since been corrected.

OUTLINE FOR REDUCING RISK

Suitability objectives for each critical issue must be reviewed. For the objectives not met to date, one must be assured that the future test program will adequately address the issues prior to the appropriate milestone. Those deficiency corrections dealing with suitability issues should be identified and testing planned to verify the correction. Each operational test should be described by the hypothesis being tested, the system configuration that will be tested, the scenario(s), and the sample size of the test (e.g., repetitions, hours, etc.).

a.   OT&E Objectives.

Are the OT&E objectives in the area of suitability listed in sufficient detail to be addressed during the next phase of testing?

This section should summarize the objectives that will be the focus of the future phases of OT&E. The discussion should be a summary of the detail that will be found in the OT&E test plan. However, it must provide enough detail to ensure that the planned test program will meet all of the management information objectives. Objectives in the suitability area should be summarized to ensure that all critical areas will be addressed.

> Objectives included evaluating the R&M of the RPV system in its operational environment and the adequacy of the planned logistics support for the system.

b.   System Configuration.

Is the configuration of the test systems to be used identified?   Are similarities and differences between the production articles and later test articles summarized?

The key to placing the test results into context is the configuration of the test articles and the realism of the test environment. The differences between the planned test systems and the production systems should be clearly identified.

> The OT Phase B will be conducted with 25 developmental sets. These sets will approximate the production configuration, except for the computer operating system. The software that will be used will have the Block II operating system instead of the Block III planned for the production units.

c.   **OT&E Events, Scope of Testing, and Scenarios.**

Is the test scenario representative of the actual support environment and are suitability issues addressed for planned test events?

Evaluating suitability requires that the system be performing at a tempo that is representative of actual operation, and that the support needs posed by the test systems be comparable to those projected for the actual operation. The summary should indicate the type of personnel who will maintain the system, the status of the logistics support (e.g., unit level maintenance, unit and direct, etc.), the maintenance documentation that will be used, and the environment under which the system is to be employed and supported during the testing. If information from outside of OT will be used by the OTA to supplement the data from this OT phase, e.g., DT data, modeling and simulations (M&S), etc., these sources also should be identified. Any planned use of M&S should be identified, along with a reference to the M&S verification plan or, in the case of existing M&S, an explanation of when and by whom they were accredited.

> The five test aircraft will be flown on representative missions of the planned duration. The systems will be exposed to the predicted combat operational stress. The sortie rate will be less than predicted (only 1 to 1.5 sorties flown each day for each aircraft). The test stations that will be available at the test site are the projected support level for a 24 UE squadron; therefore, there will be second level maintenance capability that is in excess of that planned for the operational squadron. The evaluation will be adjusted for this difference.

d.   **Test Limitations.**

Are factors identified that may preclude full and complete operational testing with concentration on suitability issues?

Limitations listed in this section should indicate which of the OT&E objectives and which COIs will not be addressed during this test phase. The limitations could include threat realism, resource availability, limited operational environments (military, climatic, etc.), limited support environment, maturity of the test system, safety, etc., that will preclude a full and completely realistic operational test. The discussion of the limitations should address the impact of the limitations on the resolution of the affected COIs and the conclusions regarding the operational effectiveness and operational suitability of the system. The COI(s) that are affected should be identified after each limitation.

> The number of available prototypes will limit the test to an examination of the performance of a single combat team. The interoperability of multiple teams will not be evaluated during this test period (COI S-6). The maintenance test equipment for the direct level of maintenance support will not be available in this test phase and the mean time to repair at the direct level will not be evaluated (COI S-2). The test site for this phase of OT does not include the variation in topology to evaluate the transmission capability under all of the terrain conditions specified in the users' need statement (COI S-8).

4.5 PART V, T&E RESOURCE SUMMARY

A summary of the suitability-related resources to be used during the test program should be provided. This should include major range and unique instrumentation requirements necessary to accomplish the OT&E suitability-related objectives. As system development progresses, test resource requirements must be reassessed and subsequent TEMP updates must reflect any changes.

AREA OF RISK

Timing and quantities of resources may not be adequate to ensure a realistic test.

The principal resource areas of operational suitability risk lie in the lack of resources to provide adequate test time or an environment that has insufficient realism. With the wide and complex array of responsibilities, test planning may fail to focus upon some of the operational suitability objectives that can be of critical importance in the operation of the system.

OUTLINE FOR REDUCING RISK

Sufficient time, test articles, and other resources must be scheduled to ensure adequate sample size and testing of the suitability characteristics of the system. Planned test article quantities, test phase duration, and other critical suitability-related test resource parameters should be identified.

a.   **Resources.**

Are the number, timing, and configuration of hardware and software test articles specified? Are unique and/or modified hardware and software test support equipments identified? Are requirements for critical operating force support and special requirements identified? Are test support spares and repair parts provided for? Have system simulation requirements been identified?

The number and timing of hardware- and software-required test articles must be identified, and be adequate to provide the required suitability test data. Differences between test and production articles should be defined, and probable impact on suitability test results should be identified. Key support equipment and technical information, required for testing in each phase and for each major type of OT&E, should be identified. Descriptions should include any appropriate measure of test duration, e.g., hours, sorties, etc.

Unique or modified support equipment should be identified by test phase. Equipment requiring special calibrations should be identified by source, and calibration requirements specified.

The operational force support requirements should be identified for each phase for testing, e.g., aircraft flight hours, ship steaming days, T&E units, etc. The support to each test element or unit and test phase should be adequate for a credible operational suitability test in light of the operation and maintenance concepts planned for the system.

Any planned operational suitability system simulations, including computer-driven simulation models and hardware-in-loop test beds, should be defined. The system simulation requirements should be compared with existing and programmed capabilities. The process used to establish the credibility of the tool should be identified.

Special data-processing equipment, special databases, and restricted/special-use air/sea/land spaces should be identified and specified. The overall test data gathering, processing, and quality control procedures should be explained and should be adequate.

> One prototype system (which is functionally identical to the production configuration) with Block I software will be dedicated to maintainability testing during Phase I. A production model with Block III software (which incorporates automatic equipment reconfiguration) will be dedicated for continued maintainability testing during Phase II. The PROSTAR model will be used in conjunction with the ALTSTAR simulation for a broader assessment of logistics supportability issues. This combination of model and simulation has been verified and validated on the Air Force SEEK ELF program using actual field data. Required test support spares and repair parts have been identified. No special equipment, databases, or test support is required for OT&E suitability testing.

b. **Budgeting and Scheduling.**

Are appropriation line numbers provided for suitability-related resources and identified by fiscal year and program element numbers? Are need dates scheduled for key suitability-related test resources?

All costs for testing should be accurately identified by program element. All items, services and/or commodities should be included. Need dates for key suitability-related test resources should be documented, including such things as unique instrumentation, support equipment, simulators, models, and test beds.

> Funding for the IOT&E is identified as follows:
>
> | | FY89 (M$) | FY90 (M$) | FY91 (M$) |
> | ------------------ | --------- | --------- | --------- |
> | RCLR-1 (PE 64321) | 350.0 | 610.5 | 240.3 |
> | RCDR-1 (PE 64789) | 132.24 | 436.2 | 670.0 |
>
> Note: Funding for specialized maintainability testing instrumentation is included in PE 64321. Associated simulation, modeling, and test bed support is included in PE 64789. Reference the program schedule for key need dates.

Chapter 5

OT&E PLAN

Test and evaluation plans are formal planning documents that provide a description of the test to be conducted and the evaluation methods. The test plan provides sufficient details about the planned test to assure the approval authority that test objectives will be addressed satisfactorily. It provides guidance to the test director regarding test execution, test approach, sample sizes, operational environment, how the threat will be portrayed, instrumentation requirements, data collection, data handling, and test data presentation during testing. It also provides the measures of effectiveness and measures of performance, as well as the comparisons to be made. The evaluation plan, which may be a separate document, describes data handling, processing, and evaluation methods.

The operational test and evaluation, and the OT&E plan, support both development and production decisions made by decisionmaking authorities during the system acquisition process. The plan details the extent to which the system's issues and criteria will be addressed during individual phases of operational testing. Operational testing should address each critical operational issue, and thereby support the evaluation of system operational effectiveness and suitability and the decisionmaking process.

There is no standard Department of Defense format for the OT&E plan. Each Service has developed its own approach to organizing the information that must be contained in the plan. Table 5-1 was constructed from the Services' policies on operational test and evaluation; it shows how the structure of this operational suitability guidance document relates to the major sections of the respective Services' operational test and evaluation plans.

This guidance document provides a process for reviewing an OT&E plan to ensure that the test program will provide for efficient and effective suitability test and evaluation. Critical Operational Issues are reviewed to identify potential implications in the operational suitability area. Once the focus of the suitability effort is determined, then the test objectives and the supporting measures of suitability are determined.

As much of the test and evaluation planning activity is directed at identifying and organizing the test assets (e.g., test systems, test ranges, supporting personnel, etc.) into planned test events and sequences, this area must be thoroughly reviewed. The structuring of the test activity should ensure that adequate testing is performed; proper structuring requires realistic testing with respect to the planned operational environments and sufficient test data to give confidence in the results. An accurate and credible evaluation of the system's suitability for operational use then can be conducted.

Table 5-1 OT&E Plan Formats

OPERATIONAL SUITABILITY GUIDE	ARMY TEP (DAP 71-3)	AIR FORCE (AFOTECR 55-1)	NAVY (OTD GUIDE)	MARINE CORPS (MCOTEA)
5.1 Description of Test Articles	1.0 Introduction	I Introduction	1.0 Introduction 2.0 Administrative Information	1.0 Introduction 2.0 System Description
5.2 Scope of Test			3.0 Scope	3.0 Operational Test - Objectives and Issues
5.3 Operational Issues		II OT&E Concept		
5.4 Test Limitations				
5.5 Test Conduct	2.0 T&E Strategy 3.0 Test Design	III Methodology IV Administration V Reporting	4.0 Operational Effectiveness 5.0 Operational Suitability 6.0 Reports 7.0 Security	4.0 Evaluation Procedure 5.0 Operational Test Conduct
5.6 Data Management		Supplement E Data Management Plan		
	Appendices	Supplements	Annexes	

90

The templates that follow address suitability considerations for each of the major content areas included in OT&E plans. These templates are organized into the following sections:

5.1	Description of Test Articles
5.2	Scope of Test
5.3	Operational Issues
5.3.1	Test Objectives
5.3.2	Suitability Parameters
5.4	Test Limitations
5.5	Test Conduct
5.5.1	Test Scenario
5.5.2	Test Hours
5.6	Data Management

In reviewing the OT&E plans from the various Services, the overriding objective is to ensure that clear linkage is provided throughout the total test program. Simply stated:

- the operational issues must be clearly defined and testing must focus on each issue;

- test limitations must be considered and appropriate adjustments made to offset the effect of the limitations;

- testing must be planned for conduct in a controlled fashion to ensure valid and accurate data are generated; and

- data must be collected in sufficient quantity and in a controlled fashion to provide substance for meaningful evaluation.

In the final analysis, all aspects of the test program must link together to provide for a credible and defendable evaluation of the weapon system's suitability and effectiveness.

5.1 DESCRIPTION OF TEST ARTICLES

The OT&E test plan coverage of the test articles should include discussion of the number of test articles and any significant differences between the test articles and the production system to be fielded. It should discuss the planned configuration and integration of the test units. Test articles will vary in their integration and maturity level during the acquisition process; therefore, testing the suitability of these systems must take these differences into consideration.

AREA OF RISK

Inadequacies in test articles can limit the ability to test and influence test results.

The planned test articles may not be completely representative of the planned production article. The number of test articles planned may not be sufficient to adequately test the system under the intended operating environments. These limitations must be identified and considered in the test planning.

OUTLINE FOR REDUCING RISK

Reducing the risk associated with OT&E testing requires that the plan clearly communicate the test assets and identified limitations and risk areas associated with these test articles.

a.   Number of Test Articles.

Is the number of test articles discussed and justified?

The number of test articles is most often a compromise between need and cost. The operational suitability implications of the test articles are primarily in the area of the statistical measures that require a number of trials or number of test hours to reach a level of confidence in the test results. The number of test articles that are proposed should be compatible with the requirements for confidence in the measures. The level of reliability to be demonstrated can be a major factor in determining the number of test articles required.

> During the system IOT&E, only five receivers will be available. Three will be deployed in the manpack configuration and two will be used in the vehicular installation. Operational suitability data will be collected to indicate any possible difference in reliability, maintainability or logistics supportability of the two different installations.

b. Configuration.

Is the configuration of the test articles comparable to the planned production configuration?

The configurations of the test articles and the production systems may differ. Differences should be reviewed and implications in the operational suitability area identified. The implications might mean that the systems will not be able to demonstrate compatibility with other items, such as test equipment, or the software will not have all of the features or diagnostics capability that is planned for the production systems.

> The test plan identifies the shortage of production isolator modules. The units to be used for the Phase IB testing will not have isolator modules of the latest configuration. Reliability data for the system will be adjusted to reflect the use of these older version modules.

c. Alternative System Configuration.

Does the test address all planned configurations of the system?

Some systems are planned for operational use in a number of different configurations or applications. The test scenario must discuss how each configuration will be tested.

> The system configurations consist of 1-channel, 2-channel, and 5-channel navigation signal radio receiver sets. The Army is the primary tester for 1- and 2-channel sets, the Air Force is the primary tester for the 5-channel airborne set, and the Navy is primary tester for the 5-channel shipboard set. Each Service's scenarios will include varying mission roles and mission types that could impact the operational suitability of the various configurations.

d. Suitability Assets.

Have adequate resources been provided for testing of the suitability elements?

The evaluation of operational suitability requires that the test be conducted with adequate operational suitability assets at the testing site. This could mean test equipment, maintenance personnel and facilities, or other items such as documentation. A limited number of assets could result in misleading data and inappropriate results.

> The second level of maintenance will consist of test equipment with the capability to verify failures and to isolate the failures to one Shop Replaceable Unit (SRU) 95 percent of the time. (It is projected that 5 percent of the Line Replaceable Units (LRUs) will have to be sent to the depot level for fault isolation and repair.) During the IOT&E Phase I, only one intermediate level test set will be available at the test site. The OT&E Test Director (TD) will determine which failed test units will be troubleshot and repaired at the test site and which are to be returned to the contractor for factory-level repair. The limitation on the number of test sets will be offset by the TD's allocation of repairable units. He will assure that a cross section of the failed units are tested and adequate suitability data are gathered at the test site.

5.2 SCOPE OF TEST

The test scope will provide a summary of the relevant information regarding the number of systems involved in the test, test location, and duration. The scope also will discuss the sequence and priority of the test phases and subtests.

AREA OF RISK

Limits and risks inherent in the planned test may be overlooked.

The scope of the test may be so limited that the system will not be exercised sufficiently to demonstrate its capability to meet the operational requirements. The description may be so abbreviated that the reader is unable to judge the limits and risks inherent in the planned test.

OUTLINE FOR REDUCING RISK

The plan should contain sufficient descriptive information about the scope of the test to indicate that the test scenario that is planned is adequate, that the test environment is representative of the intended operating environment, and that the test duration will allow operational suitability elements to be taken into consideration. Are the right factors and conditions included to ensure the system will be exercised sufficiently to capture the data needed to answer the suitability issues? The tactical context for the test should be discussed in terms of type, size of the military organizations to be represented or simulated, the operations to be conducted relative to the threat, interoperability element to be represented or simulated, and their composite relationship to the system under test.

a.   Test Concept and Scenarios.

Is there sufficient information regarding the system's different configurations and concept of employment during the testing?

The test plan should include sufficient information regarding the employment methods planned for the test and the use of any different configurations of the system. The planned scenarios should be compared with the doctrine in sufficient detail to permit examination of the test's realism. The scope of the planned test should be described, including the number of test articles, the arrangement of the test assets, the number and type of operational scenarios that will be employed, the manner of supporting the system during the testing phase, and the range and variations of test environments that will be used. See also "Test Scenario" (see section 5.5.1.a, p. 106).

> The system requires the employment of three major elements: a master station, user units, and the direct support team vehicle. Four configurations of the user units will be employed: manpack, surface vehicle, auxiliary ground unit, and airborne (fixed and rotary wing) units. The test site configuration employed will consist of up to 370 user units controlled by a master station and its alternate.

b.   Support System Concept.

Will the scope of testing exercise the support system in sufficient detail to allow evaluation of suitability issues?

The scenarios and test events should include events that will trigger the use of support resources. The scope of testing should include the use of the support structure intended to support the system once it is fielded. Any limitations within this area should be highlighted.

> Five levels of maintenance are included in the planned support concept. The test plan describes specific maintenance actions to be performed and evaluated through the third echelon.   Maintenance manuals will be evaluated for completeness and consistency with the planned maintenance skills and equipment.

c.   Test Environment.

Is the planned test environment representative of the environment in which the system will be operated when fielded?

The plan for employment during the test must be compared to the intended operational environment including the doctrine, tactics, and threat. The scope of testing should be analyzed to ensure operational suitability issues can be assessed.

> The test will involve a Marine Amphibious Brigade-size unit operating in rocky and sandy terrain during a scheduled field exercise. Weather is characterized by warm days and cold nights, with possible precipitation. Special events will include the requirement for operating in an NBC environment, wearing gas mask and protective gloves. Another event requires cold weather testing operating the system while wearing cold weather gloves, mittens, and inserts.

d.   Interoperability Issues.

Are the interactions between the system under test and other systems within the operating environment consistent with assigned missions?

The interface requirements and operating scenario of the system should be examined to ensure that required interfaces are included in the test. If the entire range of interface requirements will not be tested, are there provisions to simulate the interfaces? One should review the data requirements to ensure that information concerning interactions among systems is captured.

> The test will examine the ability of the system to effectively transfer, receive, and/or process information within the system and with external systems. The test plan includes a special interoperability test using a specified test unit, the PLRS platoon, and other personnel equipped with the EPLRS.

5.3 OPERATIONAL ISSUES

Operational issues for the system are identified in the TEMP. The list of the Critical Operational Issues (COIs) should identify operational suitability features that are critical to mission performance and the ability to place the system into field use. The issues should consider the total system, including critical subsystems and the support items, the system's wartime mission requirements, and interfaces with other systems in the operating environment. Operational suitability issues are used in developing the test objectives or test issues.

AREA OF RISK

Major issues may be overlooked.

Risks associated with the suitability issues include clarity, coverage of all missions and scenarios, and coverage of the planned operational environment. The suitability COIs may be unclear or ambiguous if they are not described completely. There is risk that the suitability testing will be focused improperly. Major issues may be overlooked. Major suitability issues related to the total operating environment may not be included.

OUTLINE FOR REDUCING RISK

In order to reduce risks associated with operational issues, the test plan must have a thorough coverage of the COIs; this includes what is operationally critical, in terms of the system, its mission, requirements, the operating environment, and the supporting organizations and structure.

a.   Critical Operational Issues.

Does the plan address the COIs that were identified in the TEMP?

The OT&E plan for the system should address the COIs identified and discussed in the TEMP. The COIs should be related to a particular phase or phases of testing for resolution. Does the OT&E plan contain test objectives to address these COIs?

> The system concept calls for two levels of maintenance for the elements of the radar. This is identified as a COI in the TEMP. The test plan has a test objective to evaluate the systems reliability and availability as it relates to the feasibility of the two-level maintenance concept.

b.   Operational Suitability Issues.

How will the suitability issues facilitate answering one or more system critical issues? Are the suitability issues necessary?

There should be consistency between the critical operational issues in the TEMP and the suitability issues and objectives addressed in the OT&E plan. The issues should contribute added focus on approaches to assessing the operational suitability. Specific test events should address each issue and provide data to determine whether the system has satisfied each issue.

> The system OT-III has the RAM operational issue, "What is the reliability, availability, and maintainability of the user equipment? The system must be sufficiently available to support the basic mission." To support this suitability issue, eleven OT&E test objectives were developed for assessment of RAM.

c.   **Operational Issue Development.**

Are the operational issues developed sufficiently to identify the operational suitability areas that should be addressed by operational testing?

The requirements documentation and employment doctrine should identify the elements of system performance that could impact the suitability of the system. From these descriptions, the suitability COIs should be identified. The test planning should relate to each COI and the rationale for selecting that particular issue. Each issue should be a focus for the test planning.

> The test plan states, "Can user equipment be effectively integrated into a wide range of weapons' platforms and function effectively in the operational environments of those platforms?" The rationale for selecting this issue states, "This COI reflects the versatility necessary for the equipment to meet the unique requirements of each of the Services involved in this program." It is apparent that there is an operational suitability issue in the area of interoperability.

d.   **Suitability Parameters. (see 5.3.2)**

Are parameters defined for each of the suitability COIs?

To determine that suitability COIs have been satisfied requires the identification of parameters that can be measured and will provide the insight required to resolve the COIs. The four Operational Test and Evaluation Agencies have published a memorandum of agreement on reliability, availability, and maintainability parameters that will be used in multi-Service OT&E. This list of parameters is an excellent starting point for identifying the parameters that need to be examined to satisfy the suitability COIs.

> The suitability parameters to be used during the test will be as follows:
>
> Reliability   Mean Time Between Operational Mission Failure
>   Mean Time Between Unscheduled Maintenance Actions
>
> Maintainability   Maximum Time to Repair (90 percentile)
>
> Availability   Operational Availability
>
> Diagnostics   Probability of Correct Detection

5.3.1 TEST OBJECTIVES

Test objectives provide an overview of what will be tested during a particular OT phase, as well as identifying the information required to evaluate whether a specific characteristic of the system meets the requirements. For each critical suitability issue there should be an objective which supports the decisionmaking process. For each test or test phase, the objective should be supported by a well understood test hypothesis. The hypothesis may not necessarily be a "statistical hypothesis." It should focus on the decision that will be made as a consequence of the test results. If the test does not influence a decision, but only provides useful data or information, then it is more properly termed an experiment.

AREA OF RISK

Faulty evaluation criteria will lead to faulty assessments.

Test objectives may be poorly developed or ambiguous. As a result, the evaluation criteria may not be properly defined, which could result in the inability to adequately evaluate test results.

OUTLINE FOR REDUCING RISK

Suitability test objectives should be well defined and descriptive of what is to be tested, as well as what data are needed to assess whether or not the objective is met. Objectives should be traceable to the COI or other issue that the objective supports. Objectives should be developed for each COI, and based on operational requirements or some other quantitative or qualitative measure.

For each test or test phase, there should be a description that includes:

- a well-defined test objective (or test hypothesis),

- the sample size planned (test hours, repetitions, etc.),

- the scenario planned for the test,

- the number of test articles,

- configuration of test articles, and

- what is missing from the system or what the differences are between the test articles and the planned operational configuration.

This description could be presented in a summary matrix.

a. Objectives Consistent with Test Rationale.

Are the suitability objectives clearly defined?

Each suitability objective should be clearly defined and support a critical operational issue. The objective and scope should be consistent, with the objective being more specific about what operational characteristic must be assessed. The hypothesis for each test should be clear from the test plan discussion. How will the test provide the data needed to prove or disprove the hypothesis? The measure of effectiveness to be used in evaluating the suitability objective also should be stated.

> An objective states "Evaluate the capability to maintain the 1-channel manpack at the organizational and intermediate levels." Organizational maintenance will be limited to built-in-test fault detection/isolation, and battery and antenna removal and replacement. At the intermediate maintenance level, there will be technicians and test equipment to perform I-level maintenance. Mean repair time will be the measure of effectiveness for on- and off-equipment maintenance. Mean repair time is the average clock hours required to return the system to a serviceable condition, excluding administrative and logistics delay times.

b. Objective Traceability.

Does each objective relate to a critical operational issue?

It is necessary to review each objective and compare these objectives to the list of critical operational issue(s) to ensure that each appropriate COI is addressed. This review should result in each objective being traceable to a critical operational issue in support of the decisionmaking process.

> The mean repair time measure of effectiveness (MOE) is traceable through the reliability and maintainability objectives and (finally) to the reliability, availability, and maintainability critical operational issue.

c. Data Requirements.

Does the objective describe what data will be collected?

Data elements necessary to evaluate the objective must be identified. The information may be collected either by test representatives or as a result of the normal course of events. How the information will be used, as well as any necessary adjustments to the data, should be discussed. (This discussion may be integrated into the OT&E plan, in an appendix, or in a separate Data Management Plan.)

> The mean time to repair includes time to access equipment, troubleshoot, repair, and check-out. Data elements required to be documented include maintenance actions, task times required to perform the repair and/or servicing of the user equipment receivers, the contributions of receiver built-in test, and intermediate test set capability to diagnose correctly the unit under test. The test team will prepare these data for input into the R&M database.

5.3.2 SUITABILITY PARAMETERS

The suitability parameters are standards by which operational suitability can be gauged and evaluated. They may be quantitative or qualitative, and measure a system's performance or a characteristic of the system that indicates how well the system performs or meets a requirement. Depending on the operational suitability element being examined, there may be multiple suitability parameters that will support the decisionmaking process, especially when the suitability element is complex (e.g., reliability or maintainability). In addition, suitability parameters used in OT&E should be representative of parameters actually used in the operating environment.

AREA OF RISK

A distorted view of the system's suitability may be presented.

Inappropriate or inadequate parameters may be selected. The appropriate number of parameters should be selected to provide decisionmakers with a complete picture of the system's suitability. In addition, the test should allow the collection of adequate data to establish confidence in the test results. In the early stages of development, the operational suitability elements may not be fully developed and the maturity level may be inadequate for operational testing.

OUTLINE FOR REDUCING RISK

Selected suitability parameters should be related to the system's operational requirements. This allows the test team to define testing that provides measures for the evaluation of operational suitability issues.

a.   **Suitability Parameters.**

Are the suitability parameters representative of those used in the actual operating environment?

The suitability parameters in the OT&E plan should be consistent with the operational suitability issues as developed in the TEMP. Multiple parameters may be required to adequately evaluate some of the suitability elements. Additional measures may be necessary for each support level. Volume I of the Operational Suitability Guide discusses suggested suitability parameters.

> To measure the maintainability of support equipment and its capability to support the mission in the tactical environment, the parameters will be mean corrective maintenance time (MCMT) at each level of maintenance, maintenance man-hours required per hour of operation (MMH/OH), and distribution of the maintenance workload and required maintenance capability at each echelon of maintenance support. The need for special tools, test equipment, and skills will be examined.

b.   Performance and Supportability Parameters.

Is there a sufficient mix of performance and supportability parameters to provide the decisionmakers with information about the total system, as well as the support environment?

The test should exercise the system in its intended missions, preferably its combat missions as well as secondary missions. The maintenance and support procedures used in the test should be representative of those that will be used when the system is fielded. The suitability parameters must be fully defined in the test plan or in reference documents (e.g., the TEMP). Failure definitions and scoring criteria also must be included in the appropriate test documents.

> One operational suitability parameter will be mean time between operational mission failures. Data collected as a result of satisfying this requirement also will be used to evaluate the mean time between corrective and preventive maintenance. In addition, maintenance publication procedures will be evaluated as to their adequacy to support preventive or corrective maintenance.

c.   Data Elements.

Have required suitability data elements been identified that will allow for the evaluation of each suitability objective?

The test plan should identify the data elements required to support the objectives, measures of effectiveness, or measures of performance.

> When measuring the mean time to repair, the data elements that will be required to be collected include date and time that the item is received at the appropriate maintenance activity, date and time that a maintenance action begins or resumes, date and time that a maintenance action stops due to administrative procedures (e.g., maintenance action ceases due to end of shift or lack of a maintenance technician), and date and time that a maintenance action is completed.

5.4 TEST LIMITATIONS

Test constraints or limitations can prevent or degrade the assessment of operational suitability and the resolution of test issues. Each constraint or limitation will result in objectives, or portions of objectives, that cannot be fully assessed. Constraints or limitations can result from the system itself, the availability of resources, environmental conditions, or the time available to conduct the operational test.

AREA OF RISK

Critical operational suitability issues may go unresolved during OT&E.

Constraints or limitations may not be identified as early as possible. Alternatives may not be addressed in test planning. As a result, critical operational suitability issues may be unresolvable during OT&E. This situation can result in systems being approved for fielding with significant potential for operational problems. Limitations must be known and understood.

OUTLINE FOR REDUCING RISK

It is not feasible from both a cost and test realism standpoint to design a test that will address every conceivable issue that should be addressed. However, once constraints and limitations have been identified, test planning can be focused to reduce the likelihood of significant voids in evaluation information.

a.   **System Configuration.**

Is the equipment to be tested representative of the planned production configuration? What will be the impact on operational suitability evaluation?

The test plan should indicate the configuration of the test articles and compare them to the planned production configuration. The system to be tested may represent only a portion of the system that is planned for eventual operation. If significant differences exist, then the potential limitations to the suitability testing should be highlighted. Differences in configuration may lead to differences in reliability levels, or an inability to evaluate other suitability elements such as support equipment or documentation.

> According to the system test plan, the space segment is not yet fully operational. Only five or six satellites may be available during short periods of the 6-hour window. A minimum of four satellites is required to obtain precise three-dimensional position fixes as required by some users. The assessments of system availability issues will be limited due to the limited total operating time.

b. Test Resources.

Are the necessary suitability test resources planned and scheduled?

The test plan and other program or test documentation should be reviewed to understand the suitability resources required to conduct the test. The personnel selected to conduct the test should be representative of the operators and maintainers of the system once it is fielded. The required level of training for operators and maintainers involved in the testing should be discussed. The personnel selected to operate the system should be representative (e.g., experience, education, and career fields) of those who will operate the system in the field.

> The test will be conducted using 10 receiver sets of the Block IA configuration. The personnel involved in the test will have received contractor-conducted "pilot" training courses that are planned for the operational users and maintainers at the first two levels of maintenance.

c. Test Scenario.

Have the appropriate geographical settings been selected for conducting the test? Where possible, have locations been selected that take advantage of natural settings, as well as weather conditions (e.g., high altitude areas with snow, as well as hot and dry locations)?

The selection of missions and test sites should be reviewed to ensure that capabilities to be tested can be supported, and environmental characteristics such as climate, terrain, and foliage are representative of the planned operational environment. This review should take into consideration funding constraints, site availability, and unique site capabilities.

> If the system is planned to be operated in geographic settings such as desert, tropical, populated, jungle, and arctic, then testing should be conducted in areas that are representative of these geographical settings.

5.5 TEST CONDUCT

The Test Conduct section of the OT&E plan should describe the general approach to be used to conduct the test, including testing for suitability. The description should include the scenario, environment, threat, tactics and doctrine to be used, and requirements to be met. It also should highlight constraints or limitations that could affect the realism of the test.

AREA OF RISK

Data that are essential for suitability evaluation may not be provided.

The planned test may not be realistic enough to provide the data that are essential for suitability evaluation. The data must be useful in evaluating the system's potential suitability for the intended operational environment. The OT&E test plan also may be of less detail than is needed to provide an understanding of the specific actions required during the conduct of the test.

OUTLINE FOR REDUCING RISK

The discussion of the suitability-related testing should be broad enough for the reader to understand the realism of the test. Test scenarios should be planned considering the operations that will be required of the system in combat, as well as the types of threat it is likely to encounter. Personnel operating and maintaining the system during testing should be representative of those who will operate the system once it is deployed. Personnel should be trained in advance of the test and be thoroughly familiar with the test plan.

a. Test Scenario. (see 5.5.1)

Are the factors and conditions of the test scenario representative of those that will be present in the actual operating environment?

The plan for operational suitability testing must take into consideration the range of environments (tactical, climatic, etc.) that the system will be exposed to when operationally employed. From an operational suitability standpoint, it is important to demonstrate the ability to support the system using the planned support structure in various operating conditions. If unrealistic test missions are used, then the suitability assessment can be optimistic or unfavorable to the system. The missions should be reviewed to ensure that the factors and conditions sufficiently portray the intended operating environment and tempo.

> The land navigation user equipment tests will be conducted along with some HELO tests. Missions will be conducted with representative operational user personnel. The aviation set will be tested in the UH-60A configuration. Test events will be conducted in accordance with the operational mode summary and mission profile of representative operational users. The support for the first two levels of maintenance will be in accordance with the planned support concept.

b.   Test Articles. (see 5.1)

Does the OT&E plan include an adequate description of the number and configuration of the test articles?

On major weapons, the test article normally consists of "worked over" prototypes; therefore, test articles are a different configuration and some are without subsystems. Test article configuration should be examined to determine its effect on the test scenario, realism, and data collection. Test articles often are early prototypes and often will not be representative of the fielded system. There are examples where hundreds of modifications have been made to test article configuration before the systems are fielded. The Staff Assistant may wish to estimate the impact of these configurations on the system's evaluation.

> Prototypes of a new aircraft have been developed by the two competing contractors. Developmental testing has been progressive and both contractors consider their prototype aircraft to be "representative" of the system to be produced. Several subsystems (including the weapon release system and the on-board diagnostics system) are being developed and are expected to be in an operational configuration in the eighth production aircraft. An aircraft with all systems completely representative of the final configuration will be available when the tenth aircraft is delivered in another three years. A DAB to approve the beyond-limited-rate initial production is planned in 18 months. (A total of 50 production aircraft are planned.) Consequently, the OT community is faced with the dilemma of a "concurrent" program structure wherein a complete operationally representative system will not be available for actual field tests before the scheduled Milestone IIIB decision.

c.   Test Hours. (see 5.5.2)

Are the planned test hours sufficient for an adequate operational suitability evaluation?

The number of test hours required to produce the necessary data to evaluate operational suitability depends directly on the system's characteristics and the mission to be performed. For high reliability systems, the number of test hours required to verify the level of reliability may be quite high. In addition, if few maintenance actions are to be performed during the OT activities, it will be difficult to provide a complete evaluation of all suitability elements. The OTA should justify the adequacy of the test hours available for the OT.

> The number of test hours scheduled for the system OT was limited due to restrictions on test funds and limited spare parts. Had the OT been conducted as planned, the spares and test hours may have been sufficient; however, because there was a test schedule slip, the hours were not sufficient to support the prolonged test schedule. The Test Director made a determined effort to execute the test as planned. He was willing to except test data, although some were of limited value. Hours became the main driving force in the test, compelling the Test Director to except a lot of unusable data. The SA should ensure that test hours, as with any test resource, are not the driving force in test execution. He should ensure that data collection, test realism, and, only then, schedules are the primary considerations in the OT&E.

A test scenario is a set of circumstances by which a system is tested in a representation of its intended operational and support environment. The test scenario should be based on mission scenarios, support concept, critical operational issues, objectives, and test limitations. Consideration must be given to terrain, weather, and other operational factors.

AREA OF RISK

The test scenario may fail to reveal operational suitability deficiencies.

The operational test scenario may lack the detail and realism needed to reflect the intended operational use of the system. In addition, it may not provide for adequate operation of the total system to exercise realistic demands on the support system.

OUTLINE FOR REDUCING RISK

Scenarios should be based on realistic factors and conditions that will be present in the intended operating environment. Mission roles and types should be used and varied in a manner that will ensure that test objectives and, ultimately, the critical suitability issues can be answered with an acceptable degree of confidence.

a.   **Test Scenario.**

Is the test scenario representative of the planned operating scenario and broad enough to allow for sufficient data collection to analyze suitability issues?

The degree of realism of the test scenario and mission determines the extent to which the support system and operational suitability can be evaluated. If the missions used are unrealistically severe, then suitability may appear to be less than it really is. If the missions are less severe than the expected operational missions, then the evaluation of operational suitability elements may be optimistic compared to what is achievable in the operational environment.

> The system tests will be conducted with a cadre of command and control personnel, equipment, and sufficient user units and operators to bring the community up to 370 user units. There will be no opposing force except for the EW jammers during this period. The scenario calls for two days of scripted defense and two days of scripted attack over approximately the same area as the Phase I test. Tests will be conducted 12-hours a day, which will allow data processing overnight prior to continuing the next day of testing.

b.   **Test Events.**

Do the test events represent the minimum number of mission types that can be expected to be performed with the system? Are affected operating activities identified and test events planned for each?

The planned testing should include a representative set of the planned missions. There should be planned test events for all system functional elements; they should involve all mission types. The events must be representative to ensure that suitability can be evaluated.

> As a minimum, 16 corridor and six linkup aviation mission types, under nap-of-the-earth benign conditions, will be accomplished using nine pre-surveyed waypoints. In addition, a minimum of six corridor and six linkup mission types, under nap-of-the-earth electromagnetic conditions using nine pre-surveyed waypoints, will be accomplished. All operational missions will include en-route navigation, present positioning data, and termination fixes.

c.   Maintenance Concept.

Is the maintenance and support concept to be used during the testing an accurate representation of the operational environment?

Maintenance support to be provided during testing should be representative of how maintenance will be performed after the system is fielded. Operational suitability elements (e.g., test equipment, maintenance documentation, and man-machine interfaces) that impact the effectiveness and efficiency of maintenance operations should be included in the evaluation.

> The system has been designed to use the established five echelons of maintenance support. During testing, first echelon maintenance will be performed by the radio operator for all ground configurations and by the organizational maintenance activity of each user of the airborne radio set. Second echelon maintenance will be performed by the Electronics Maintenance Company and will consist of operating and performing checks to determine fault areas prior to evacuating to third echelon maintenance. Third echelon maintenance shall be performed by the contractor and will not be evaluated in this phase.

d.   Test Environment.

Were all potential operating environments considered in the construction of the test environments?

The planned operating environment can be described in many ways, including weather and geographical conditions, electromagnetic conditions, and battlefield conditions including smoke, noise, and CBR. The operational testing usually is limited in the range of environments that can be addressed. Therefore, it is important to assure that those conditions that potentially could have adverse impacts on the operational suitability of the system are addressed. Adequate consideration of these factors should be included in the test scenarios.

> The Air Force testing of the GPS UE requires support from the Air Force Electronic Warfare Center to test the susceptibility of the 5-channel airborne set to jamming of the satellite downlink signal. The Air Force will provide the technical expertise on the selection and operation of an airborne jammer. The jammer will be operated in accordance with Warsaw Pact radio electronic combat doctrine during six bombing missions, and three level and three loft profiles.

5.5.2 TEST HOURS

The number of test hours planned for the system under test is directly influenced by the mission types and roles that the system must perform. The length of time required to complete a mission plus the number of missions in the scenario should be representative of the mission lengths and duration that will be required in the intended operating environment.

AREA OF RISK

Operating time will not provide representative data and thus will lead to faulty conclusions.

The number of test hours or the mission duration may not be adequate to provide the necessary data for evaluation. The amount of test data that is available for evaluation is tied directly to the number of test hours. To provide the decisionmaking authority with a realistic assessment of the operational suitability of the system, careful consideration must be given to mission length and duration.

OUTLINE FOR REDUCING RISK

The number of test hours should be sufficient to demonstrate the system's operational and physical characteristics that can impact the operational suitability of the system. The test hours should take into consideration the impact that a short test will have on statistical confidence, as well as the long-term effects on operational suitability.

a.   Hours to Demonstrate Characteristics.

Is the number of test hours sufficient to demonstrate the operational suitability characteristics?

Operational suitability characteristics, such as reliability and maintainability, require relatively lengthy test periods to adequately demonstrate their achieved levels. In addition, other operational suitability items, such as technical data, have many alternatives, sections, or paths that cannot be exercised in a limited period of operational testing. The test plan must deal with the need for lengthy test periods and realistic levels of test resources.

> A limited number of test hours will be available due to small numbers of host vehicles, limited satellite coverage, and a compressed test schedule. To supplement the operational testing data, additional data will be available from a number of platforms not directly involved with this dedicated OT&E. This additional operating and failure data will be used to expand the system suitability database. Careful scrutiny and evaluation of the supplemental test data will increase the confidence level of the suitability results.

b.   Long Term Effects.

Is the amount of time dedicated to testing individual test articles sufficient to evaluate the long-term effects on the articles?

Test items may have risk areas that require knowledge of the effects of lengthy periods of operation. Although statistical methods sometimes allow the testing of a large number of articles for relatively short periods of time, there is still a possibility that unknown risks will be realized when the system is exposed to longer periods of operation. When the OT&E plan calls for short periods of testing, the effects and risks associated with use of the system for longer periods of time must be considered.

> A selected number of the test articles from the OT&E Phase IA will be provided to support the test. The use condition will be similar to that employed in Phase IA. Failure data from this extended period of operational use will be evaluated to determine if any long-term failure modes are likely to exist. The results of this additional operation will be provided as an appendix to the OT&E report.

c.   Statistical Confidence.

Is the test time sufficient to ensure any required test confidence?

The system's operational requirements for reliability and maintainability may include the need for confidence levels. Confidence levels may be required also as part of the reporting requirements. The length of the operational test should be reviewed to assure that the required level of confidence can be achieved. The plan should state any requirement for confidence levels to be calculated for operational suitability measures. DoD 3235.1-H may be used as a reference in assessing the confidence levels and test times included in the OT&E test plan.

> The mission reliability of the system shall be measured in terms of the mean time between operational mission failures (MTBOMF). The point estimate MTBOMF that results from the scoring conference data shall be used to compute an 80-percent lower one-sided confidence interval. The 80-percent confidence value shall exceed the requirement stated in the approved ROC. Operational mission failure will be determined by a formal scoring conference in accordance with the approved failure definition and scoring criteria.

This section describes, in general terms, how the testing organization plans to collect, organize, reduce, verify, manage, control, analyze, and store the data needed to perform the evaluation. Suitability evaluation requires data on many aspects of the system's operation and on the support elements. The data management plan must provide for the proper collection and control of the test data.

AREA OF RISK

Inappropriate or incorrect data may be used for critical evaluations.

Poor data management can result in ineffective and inefficient data collection, control and reduction, and data evaluation methods that are unable to support the decisionmaking process.

OUTLINE FOR REDUCING RISK

The suitability data gathered during testing must be properly managed to ensure credibility and validity of the conclusions about the system being tested. For determining the proper types and amounts of suitability data to be collected, and the way in which the data are controlled and evaluated, a well defined data management plan is required. This plan may be part of, or a supplement to, the test plan.

a.   Quality Control.

Does the data management plan include quality control methods to ensure incorrect or inaccurate suitability data are eliminated?

The data management plan should make provisions for tracking the data from initial receipt until final input into the master database. The plan should include procedures for collection forms, data verification, data reduction, and data entry and editing.

> When suitability data discrepancies are discovered through regular quality control checks of the data, the data will be returned to a previous step in the data flow process for resolution. Ultimate control of the data is the responsibility of the quality assurance test manager; however, the day-to-day quality control of the RAM data is the responsibility of the RAM data manager. Data collection personnel will perform quality control checks to identify missing or logically incorrect entries. Quality control during the data reduction process will be accomplished by both manual and automated methods. Data managers will ensure data reduction procedures are consistent across missions and that results are checked by someone other than the person doing the actual data reduction. Scrubbing, correlation, and merging of automated data from various sources with other data, both manual and automated, will be controlled by computer programs, when possible. The program(s) will be verified to be functionally correct prior to the pilot test. Data entry and edit routines will be controlled and monitored by computer programs. No data will be loaded into the master database until all data entry errors identified by quality control procedures have been corrected.

b.   Data Collection and Validation.

Have procedures and responsibilities been established for manual and automated suitability data collection? Does the test plan or data management plan describe the forms that will be used in the data collection process?

The test plan or data management plan should include procedures to control the data collection process for both manual and automated data. These procedures should address how the data are to be organized, reduced, verified, managed, controlled, and stored. The process to include or exclude data from the database should be described. How will "no tests" be identified? The data collection forms should have been developed and approved, and the responsibilities for data collection defined. Test personnel should be trained in manual and automatic data collection prior to the start of the test. Sections 6.2.c and 6.7 also discuss data collection and validation.

> A data collector will be assigned to each of the systems under test for the purpose of RAM data collection and to track equipment operating time and events at the time of a RAM incident. RAM data forms will be based on forms from the RAMES management information system, but tailored for use on the system. The RAM data forms also will be used to collect RAM data on the MX-379 test set and other support equipment. RAM forms will be supplemented by narrative commentary on the acceptability of the technical documentation and special tools used for each maintenance action.

c.   Data Processing and Analysis.

Are data processing procedures described in sufficient detail to allow assessment of the procedures?

The planning should include data handling procedures for both manual and automated suitability data. The procedures should include the necessary quality control for the data flow from the data collectors until it is finally entered into the master test or suitability database. Procedures for analyzing suitability data should be described in sufficient detail to provide an understanding of how the data will be reduced and used in critical evaluation. Equations and algorithms should be addressed.

> Data management personnel will log all manual suitability data forms provided by the data collector, review them for consistency and completeness, and add any additional information that may be needed for internal control. If there are problems, the forms will be returned to the data collection section for resolution. Once the forms are determined to be correct, the data will be entered into the microcomputer database. Automated quality control and edit routines will ensure data validity. The quicklook performance and RAM reports will be generated for analysis and validation by data managers.

Chapter 6

OT&E OBSERVATION

In addition to reviewing test documentation such as the TEMPs, OT&E plans, and OT&E reports, the DOT&E Staff Assistants have an important responsibility in performing on-site observation of the actual conduct of the operational testing. (Title 10, U.S. Code, Section 138 states that the DOT&E may require that observers who he designates be present during the preparation for and the conduct of the testing part of a DoD OT&E.)

Test program reviews for programs assigned to an individual Staff Assistant should be planned to ensure that appropriate priority is assigned for the coverage required for specific tests. Considerations include which test events require a DOT&E presence, and the degree of on-site monitoring which will be required. Given that a particular event is to be covered, the level of involvement may vary considerably. The Staff Assistant may arrange for a team of DOT&E observers to be on-site for the duration of the test, or there may be only one DOT&E observer on-site during only one or more of the critical testing periods.

Once the particular test event is identified and the degree of DOT&E coverage has been decided, preparation for the test observation is required. This chapter provides guidance to assist in preparing for on-site visits, and it provides information on items to be considered at the test site. At the site, there are general areas that need to be considered. The management of the test site should be examined, and specific items in the test events should be compared with the discussions in the test plan. Finally, documentation that is used during the operational test events, personnel involved in the operation and maintenance of the system under test, the data collection program, and the test scenarios also must be considered.

This material was prepared within the context of operational suitability and has as its focus the oversight of areas of testing that impact on the operational suitability results. However, most of the items in this chapter apply equally well to the testing for the operational effectiveness objectives.

The templates that follow address suitability considerations for a number of the activities that are involved in observing the operational testing. These suitability considerations are:

6.1	Planning for Test Site Visits
6.2	General Observations at the Site
6.3	Test Site Management
6.4	Comparison to the Test Plan
6.5	Documentation
6.6	Test Personnel
6.7	Test Data
6.8	Test Scenarios

6.1 PLANNING FOR TEST SITE VISITS

This section discusses considerations for the general "test observation" aspects of the OT&E oversight responsibilities. Observation of the actual operational test activities is needed to provide for credible analyses in DOT&E-prepared beyond-LRIP, annual and other DOT&E reports. It also provides first-hand knowledge to support other DOT&E responsibilities, such as review/concurrence on the test and evaluation section of the Congressional Data Sheets and Selected Acquisition Reports (SARs).

AREA OF RISK

Poor preparation for on-site visits may result in invalid conclusions.

A prime contributor to this risk concerns the balance between the amount of time dedicated to day-to-day Pentagon responsibilities and time expended in the field. The ability to ensure accurate and credible DOT&E reports can be strengthened by first-hand observation of test events. For the on-site visit and observation of the test activities to be effective, preparation must be made prior to arrival at the test site.

OUTLINE FOR REDUCING RISK

a.   **Awareness of Current OT Schedule.**

Are current test activity schedules available in the DOT&E office?

Test schedules in the TEMP as well as detailed schedules for discrete test events are subject to long- and short-term revisions. Attendance at selected Test Plan Working Groups (TPWG) or Test Integration Working Groups (TIWG) will provide valuable insight into schedule variability, the status of system development, and the readiness for operational test activity. The Program Manager, the OTA, or the Service Staff's Program Element Monitor are sources of updated activity schedules. Informal dialogue with members of the test team will facilitate an ongoing cognizance of the program status, at times prior to notification from the Service staff. Schedule awareness should include significant pretest activity, e.g., Test Readiness Reviews, in addition to actual test conduct. DOT&E staff members can leverage their effectiveness by utilizing all sources of information including the operational test agencies, program offices, and using commands. These groups assist in establishing test objectives, identifying needed resources, providing forums for discussion and resolution of test and integration subjects/problems, and related activities. During attendance at meetings, there is the opportunity to gain better insight into underlying issues and the associated potential impacts to the operational suitability of the weapon system under test.

> Test schedules were changed due to the late arrival of the test article, time required to install and calibrate instrumentation, and time required to conduct pretests and training. After two weeks of test trials, the Data Authentication Group (DAG) recommended that early trials be rerun due to the poor data from those events. At this point, the test was 4-1/2 weeks behind the original schedule.

b. Structured Planning for Each Test-Observation Trip.

Will the limited on-site time be used effectively by the test monitors?

When planning for a trip to observe test operations, an activity checklist is extremely useful. This checklist will assist in test observation activities and provide reminders of critical issues or risks identified in key source documents as well as those surfaced in prior OSD reviews and congressional interest items. This approach ensures a consistency of purpose and maximizes the benefit of the on-site activity. Field observations will include deviations from planned activities and/or discrepancies that could impact quality of test results.

The Staff Assistant prepared a checklist of pertinent information prior to the scheduled visit to the test sites:

Configurations	Interagency Coordination
Hardware	Lessons Learned
Software	Analysis Procedures
Critical Operational Issues	Management Plans
Test Design/Methods	Policy Documents
Data Management	Security Requirements
Environmental Considerations	Simulation Certification

---

c. Preparation for Accurate On-site Observation Reporting.

Will the DOT&E observers recall and accurately report what actually happened during the monitored activity?

An accurate account of the monitored activity is crucial. Careful notes must be taken to assure that incorrect information or conflicting information is not reported. In some cases, it may be desirable to use a portable tape recorder/dictation machine as the test event progresses; this would be supplementary to voice-tapes (which may be recording radio transmissions, etc.) used by test participants. Particular attention should be given to recording the specifics of unusual events and those which caused deviation from planned scenarios. Test monitors should maintain a log, recording the conditions which existed during the events, or scenarios.

The DOT&E Staff Assistant used a hand-held cassette recorder to record her observations during the testing. These tapes were reviewed each evening for major significant points to be included in the trip report. Upon return to the Pentagon, portions of the recordings were transcribed completely by the DOT&E administrative support staff. This method of recording test site observations proved to be more thorough than notes or personal memory.

6.2 GENERAL OBSERVATIONS AT THE SITE

The benefits of personal observations at the operational test site are many. The Staff Assistant (SA) may discover information that would not have been thought important enough to be included in the OT&E report, but that will be important in the deliberations on the meaning of the OT&E results. The SA must be sensitive to this type of information before going to the test site, and have a sense of priorities about what observations and conclusions are important.

AREA OF RISK

Unexpected items or unusual events may yield test results that differ from those expected when the test was planned.

Once the operational testing is initiated, unplanned situations will occur and the testing that was intended by the people who reviewed the plan may not be accomplished. This situation is due in part to the fact that operational testing can never be completely described in the written document. The readers of the test plan information all will bring different interpretations to what the written word means. Once the test begins, it is common for these differences to become known. When the DOT&E Staff Assistant visits the test site, it is essential to identify any significant difference between what was "thought to be" and "what actually is." This identification is not for the purpose of redirecting the testing, but to place critical test results within the proper context.

OUTLINE FOR REDUCING RISK

a.   **Monitors Must Watch for "Good Intentions" Deviations.**

Are deviations from approved test documents occurring inadvertently or through overzealous efforts of test participants?

A part of the DOT&E responsibilities is to designate observers for the planning for, and conduct of, the testing portion of OT&E. Experience in participation in test reviews, readiness briefings, post-test debriefings or reviews also can be key elements of DOT&E monitoring activity. Prior experience can reveal that deviations have been highlighted during internal DoD reviews, as well as by agencies such as the General Accounting Office (GAO). DOT&E representation at test-related activities can identify impending or actual variances so that appropriate corrections can be completed, thereby ensuring credibility of the test programs. During the on-site visit, the SA should search out deviations from the test plan caused by instrumentation, personnel problems, or schedule changes that may effect the validity of the test results.

> During an operational test, the Engagement Line-of-Site System became inoperative and estimates for making repairs in a reasonable time were not good. Due to this failure, the Test Director elected not to run additional trial events of the system's ability to detect and track the threat system, even though the system's ability to detect and engage systems was a COI that was to be addressed during the test. The Test Director stated that data would be obtained from models and simulations to answer these issues.

b.   Resolution of Uncertainty and Trip Report Preparation.

Are there "observations" that the Service Test Director will not agree with from an accuracy viewpoint?

Observations may be reviewed with the Service Test Director to assure that the test monitor has a complete understanding of the actual circumstances and events. However, test monitors will not allow such review to impact reporting of their observations or their consideration in subsequent DOT&E analyses.   Upon completion of the monitoring of a specific test event, the Staff Assistant (SA) should coordinate with others who monitored other aspects of the same testing, and all significant observations should be consolidated. Prior to departing the test location, the SA should review his major observations with the Service Test Director, and any resultant con-flicting conclusions should be highlighted in the SA's trip report. A database of significant findings from the monitored activities should be compiled by the SA and maintained for use in development of B-LRIP reports, annual reports, etc.

> From interviews at the test site, the SA determined that the mission pilots were flying practice "safety" sorties over the planned flight profiles on the day prior to the OT test events. These practice flights would have skewed the test data. When the observation was discussed with the Test Director, it was discovered that while the "safety" flights were being made, the flights did not involve the pilots assigned to the OT test events.

c.   Data Authentication and Validation.

Are data being validated and authenticated by an independent group?

When required by the test complexity or data volume, a Data Authentication Group (DAG) should be formed to perform data evaluation and certification of engineering analysis. Normally, the DAG is independent of the data management and quality control process and is not under the supervision of the data manager; it provides a level of quality assurance above that to be expected from the data management quality control function. Each DAG should be tailored to the unique requirements of the test. There should be a procedure that includes the DAG's con-cept of operations, organization, responsibilities, validation procedures, reports to be produced, and required schedules.

> There are critical windows in the data collection effort. Should the windows be passed without appropriate actions being taken by the test team, valuable data are lost forever. During the test of an aircraft, it was determined that much of the data collected from the MIL-STD-1553 data bus were of very limited value because the on-board recorders and playback systems had not been developed sufficiently to allow access to the event data. If the Test Director/chief of instrumentation/chief of data management had the playback machines developed and tested in sufficient time to be used to evaluate the data, valuable information could have been gained and test resources preserved.

6.3 TEST SITE MANAGEMENT

The proper management and control of the test site and the test site activities are essential to providing the discipline necessary for an effective operational test program.

AREA OF RISK

Poor management procedures or procedures that are not followed can lead to invalid test results.

Converting the OT&E plan into specific actions requires test site management and operating procedures. The Staff Assistant should assure that planning has been performed for adequate management of the test site and that procedures have been issued that formalize the planning.

OUTLINE FOR REDUCING RISK

a.   **Program Documents.**

Have the pertinent program test documents been analyzed?

A prerequisite to effective monitoring of system testing, simulation activity, demonstrations, or other events is an in-depth understanding of existing guidance related to the event (derived from the TEMP, Operational Test Plan, and other documents). Pretest analyses assist in clarifying the important test parameters, developing linkages to tie field test data with modeling/simulation data, and in establishing the analytical structure to perform final analyses. An early item would be verifying OSD approval of the TEMP as well as the DOT&E approval of the operational testing of the system.

> Prior to leaving for the test site, the SA made copies of the pages from the unclassified test plan that describe the test events that were planned. The TEMP for the system had been approved six months earlier and agreed with the OT&E test plan that was reviewed three months earlier. Major suitability deficiencies from the previous test phase were summarized as low percentage of fault detection by the diagnostics subsystem, the immaturity of the maintenance software, and very high mean times to repair. The corrections to these deficiencies were to be verified during the test events to be observed.

b.   **Instrumentation Plans.**

Have instrumentation plans been analyzed to determine their effect on the tactical scenario?

A prerequisite to effective monitoring of a system test is an understanding of the instrumentation that will be used to collect the data required to evaluate the system. If not carefully planned and managed, instrumentation will tend to "drive" the test. After review of the instrumentation plan, the SA should be satisfied with the proper balance between test realism and the requirements for instrumentation.

> During a helicopter OT&E, there was a requirement for instrumentation to be calibrated before each test event. The calibration was accomplished by flying the helicopter close to the instruments within minutes of the start of each trial. This requirement provided a tactical advantage to the tested system's crew because they had an opportunity to reconnoiter the battlefield and it reduced the tactical realism of the test. As it turned out, the instrumentation that required this calibration was not required for the test and the events were evaluated without the use of these data.

c.   **Data Management Plans.**

Have data management plans been analyzed to determine if the plans conflict with test realism?

Reliable operational testing requires test realism. Data collection may tend to conflict with test realism; therefore, there must be a compromise which maximizes the combined issue of realistic testing and thorough data collection. The SA should ensure that data collection and instrumentation is not causing the test to be unrealistic. The goal of instrumented data collection should be clearly specified in the data management plan. Although a realistic percentage of the total data generated must be developed to ensure proper analysis can take place, as a minimum the plan should identify the critical items for collection, to include a minimum quantity of each. In some cases the test event may be a singular event, in which case it becomes a critical item to collect these data and redundancy should be planned to ensure adequate collection.

> During the operational test of an aircraft, the data collection plan stated the requirement for an on-board tape recording system to capture MIL STD 1553 Databus information. The recorders required the crew to land aircraft every 25 minutes to change the tape in the recorder. The test trial scenarios were developed around this data collection requirement, which made it very difficult to execute the test trials and collect meaningful data from other data points. Test realism also was limited because of this data collection requirement.

d.   **Data Authentication Group.**

Is there a group of professionals established to validate data?

The Test Director or Program Manager often will establish a Data Authentication Group (DAG) to verify and validate test data, in addition to assisting in data reduction, quality control, and the identifying anomalies in the system, instrumentation, and test data. The DAG must be independent of the system developer and data manager and should report directly to the Test Director. It is important that a standard operating procedure (SOP) be written for the DAG and that this SOP be closely followed.

> During the second week of the vehicle test, the Data Authentication Group announced that the night navigation trials data that already had been collected were sufficient to provide a "high level of confidence" evaluation of the system's performance. The Test Director elected to terminate the scheduled additional night navigation trials and use those resources in another area where confidence levels in the data were not as high.

6.4 COMPARISON WITH THE TEST PLAN

Owing to a host of reasons, the actual test will differ from the planned test. As with most plans, test plan usefulness ends with plan execution. However, test planning documentation has an added value to the test community in that it serves as a check sheet for the operational tester during execution phase. The SA should determine the degree of variation from the plans that the tester has been required to take in order to conduct the test. These unplanned variations often are weak links in the process and should be evaluated carefully.

AREA OF RISK

Actual test is different from planned activities.

During the actual test conduct, there may be room for on-the-scene decisions that result in the test being significantly different from that envisioned when the test plan was written and approved. These undefined areas need to be examined during test site visits and an understanding gained of how the items will affect the resulting test data and evaluation.

OUTLINE FOR REDUCING RISK

a.   **System Configuration.**

Is the system configuration to be tested the same as that identified in the test plan?

In some situations, the system configuration planned for use during OT either is not available or it is delivered to the test site with some unexpected deviation. In limited cases, there is a deviation in the number of test articles. The limited assets available early in the development program may result in a different number of test assets being available at the test site from the number that was planned. The Staff Assistant should determine what the system configuration deviations are and evaluate their impact on test results.

> The system configuration of the three prototype helicopters provided for the OT were different from one another and significantly different from the planned fielded system. Maintenance personnel training was difficult due to these differences. (More than 1600 modifications were made to the prototypes before a fielded system was produced.) Significant contractor support was required to maintain the systems during the test. Many prototype subsystems were of such early configuration that their performance was difficult to measure.

b.   **Test Limitations.**

Will test limitations significantly impact the test results?

The SA can review all listed test limitations to ensure that they do not impact on the ability of the test to meet the stated objectives. The test limitations should be clearly explained and it should be determined if the limitations can be avoided or resolved by the responsible organization.

> The testing of a combat earth mover was limited by the types of soil and weather conditions to which the system would be exposed at the selected OT location. After evaluating the system's performance at the test site (clay/sandy soil), it was determined that it would not be possible to estimate the system's earth-moving capabilities in rocky and other types of soil. Additional OT at other test sites was required to provide the necessary evaluation report.

c. Unusual Pretest Procedures.

Are the test articles or support equipment subjected to unusual pretest maintenance?

During most OT periods, there are test events spaced throughout the testing period with periods of minimal activity in between. During the on-site observation, the range of activities that occur during the nontest periods should be assessed. Are the test systems subjected to unusual pretest maintenance? Are the test articles to be included in the testing selected from a "pool" of available assets? If so, is the selection representative of what the operational commander would do, or is the selection likely to skew the resulting test data? While there may be motivation to maximize the use of scarce test-range time, the test team must guard against the use of unusual pretest maintenance that skews the reliability or maintainability data.

> During the operational test and evaluation of a mortar system, it was determined that the ammunition casings had been machined rather than cast, as is normally the case. By machining the casings, the weapon proved to be very accurate although the ammunition was very expensive. The mortar is an area weapon and the normal method of manufacturing ammunition is by casting. When cast ammunition was obtained and fired, the mortar did not meet accuracy or dispersion pattern requirements at extended ranges.

d. Maintenance Activities Reflect Operational Concepts.

Is troubleshooting or system repair activity performed by representative personnel?

In the case of a major defense acquisition program, no person employed by the contractor for the system being tested may be involved in the conduct of the (initial) OT&E that will satisfy the B-LRIP reporting requirements of Public Law. In addition to pretest maintenance activities, post-test involvement of such contractors could be a cause for concern (an example being if contractor personnel became overzealous in "Scoring Conferences" by defending the cause/chargeability of a system anomaly).

> During the operational pretest, the system's contractor monitored the missile system's failures and made corrections on spare computer circuit boards. Before the start of the OT, the contractor inserted the modified boards into the test articles. The substitution of modified circuit boards disabled all on-board test instrumentation and caused four missile shots to be made without data being collected. The uncoordinated modification of the test articles during pretest caused extensive delay and cost to the test program.

6.5 DOCUMENTATION

The documentation at the test site should include the operator and maintenance instructions and various supporting manuals, as well as manuals relating to the system software. The role of the documentation is to provide for system operation and maintenance that is consistent and that can serve as a basis for evaluation of the system.

AREA OF RISK

System documentation is not evaluated, nor does it provide the foundation for realistic system operation or maintenance.

When the operational testers are not using the intended system documentation for operation or maintenance procedures, then there is little assurance that the data from such a test are representative of what will occur in operational use. The system documentation should provide a foundation of system operation that renders the test data usable in operational evaluation. On-site observation of testing should include an effort to assure that the documentation is being properly used by the test site personnel.

OUTLINE FOR REDUCING RISK

a.   **Source and Status of Documentation.**

Is system operation and maintenance documentation in a mature state to allow use during testing?

System operation and maintenance documentation normally is developed either by the system's developing contractor or another contractor. If development is not begun early, the documentation usually is in draft stage at the beginning of OT&E. Often document developers will attempt to modify the documents during pretests and tester training. These modifications usually result in significant changes in the system's operations or maintenance. The SA should review the system's operations and maintenance documents and determine their state of readiness for the OT&E; this review should include the assessment of the effect of early drafts and/or poorly presented and partial documents on the ability of the crew and maintenance personnel to perform during the test.

> During the training phase for a weapon system test, it was determined that operation and maintenance documents were in an early state of development. The Test Director requested the system's contractor to extend the crew and maintenance training to compensate for this lack of documentation; the draft documents were used in training and during the operational test. During the test, many crew/maintenance errors could not be classified as to whether they were caused by improper training, poor reference documentation, or personnel errors.

b.   Use of Maintenance and Operating Documents.

Are test players using the operational and maintenance documents to assist in the performance of their duties?

A good indication of the level of completion and the confidence that the troops have in the documentation is their decision to use the documents during the test. When it is determined that the documents are not being used, it may be desirable to ask the test player's opinion about the documents. When the documents are available, the SA should determine if they are presented as being complete, and if they are considered accurate by the test personnel.

Just prior to the operational test, it was determined that the system's maintenance documentation was in an early draft stage. The Training/Logistics Command personnel were planning to use the OT&E to evaluate the maintenance documentation and make the necessary changes before going back to the contractor with final revisions prior to publication. The personnel had developed sampling techniques such as questionnaires, survey forms, and observation techniques to be used during the OT&E. The Staff Assistants should determine if other segments of the Service (i.e., training, logistics, doctrine, or tactics) are using the OT to gather information and if their methods might adversely affect the test realism, data collection/evaluation, or test operations. He also should determine the status of the system's manuals prior to the start of OT.

6.6 TEST PERSONNEL

Observation of the operational test gives the Staff Assistant a sense of the skill level and qualifications of the personnel. Direct contact with members of the units or test team also will provide information about how representative of typical user troops the operations and maintenance personnel are.

AREA OF RISK

The use of unrepresentative personnel may result in invalid data.

If the personnel who are operating the system during OT are more highly skilled than will be the planned operational troops, the system will perform better than might be expected in the operational units. Likewise, if the maintenance personnel are unusually highly skilled, the suitability of the system may be optimistically reported. In the visit to the test site, the Staff Assistant needs to ascertain if the skill levels involved will affect the test data.

OUTLINE FOR REDUCING RISK

a.   **Tester Training.**

Are test personal properly trained?

Training plans and certification plans for test personnel should be developed and published early in the full-scale development phase. Errors by test personnel usually are expensive and often cloud the reason for test failure. The SA should identify the training performed prior to the start of test and determine if additional training was given to test personnel as a result of the pretest events.

> During the pretest trials for the OT of a helicopter, it was determined that the test system crews had received additional training on aircraft subsystems, whereas the crews for the baseline system had not received this additional training. The subsystems involved basically were the same; therefore it was decided that the baseline crews should receive the same blocks of instruction as did the test system's crews. When comparing a new system with a baseline system, training becomes very important and must be carefully monitored.

b.   **Use of Contractors.**

Are the system's developing contractors participating in the OT&E?

Personnel from the developing contractor are prohibited from participating in the conduct of operational testing. During very early test phases, the contractor may have a role in some of the suitability areas, e.g., performing second or third level maintenance, or assisting in maintenance that is specific to the developmental systems. If there is some level of contractor involvement, the Staff Assistant must examine the way in which this involvement might be influencing the OT data results. Are the Service maintenance personnel "consulting" with the contractor during routine maintenance?

During the OT of an aircraft, the contractor and subcontractors were allowed to move into a motor pool complex near the test site. It was very difficult to keep the contractors away from the test systems. On one occasion, a contractor was observed pushing a military maintenance person out of the way so that he could make adjustments on the aircraft engine.

c.   Use of "Golden Crews."

Are the crews involved in the OT typical of those who would be expected to operate and maintain the system once fielded?

If the skill levels of personnel operating or maintaining the system in the OT are higher than those of the crews who will operate or maintain the system once it is fielded, the OT data and the evaluation results can be skewed. The term "golden crews" is used to identify which of the operating/maintenance personnel have skill levels higher than those of the field crews. If the skill levels of the field crews are primarily mid-level, not all the test maintenance/operating personnel should be highly skilled. Is the skill level being used of such a high degree that it skews the OT data and the evaluation results?

An aircraft test team was formed with maintainers fresh from the operating units of predecessor systems. While at the test site, the SA determined that while these troops were representative of the operating units, over 75 percent of them were assigned the 7-level skill code. In the operating units, only 25-30 percent of the personnel are 7-levels. The evaluation of the OT maintenance data must consider that some of the maintenance actions that were performed by 7-level personnel would be performed by 5-levels in the operating units.

d.   Stress Level of Personnel During Test.

Will the test crews be placed under the operational stress that would be expected under combat conditions?

The stress levels that are present during the operational use of combat systems cannot be duplicated during operational testing. While this statement is true, it should not be taken as a rationale for avoiding the replication of realism--enemy fire can be simulated and unexpected events can be entered into "free-play" portions of tests. The human factors evaluation of stress levels in combat requires that consideration be given to all of the items that are unknowns during the actual combat use of the system.

The troops used in the OT of a shoulder-fired missile were being evaluated for human factors, portability of the missile, firing effects, ability to guide the missile, etc. The crews selected for this portion of the evaluation should not have more experience firing the predecessor system than will the typical user troops. If the user troops only have experience with missile simulators, then the crews used in the OT should only have an experience level that is representative. Stress can be introduced into the OT by having the user troops operate for long periods, similar to the projected operational use, and at the projected pace, or tempo.

6.7 TEST DATA

The observation of the operational test site provides the Staff Assistant with an understanding of the test organization's approach to data collection, reduction, and reporting. The data collection plan will include the methods for collecting, marking, handling, storage, and disposition of test data, as well as the training required for the data collectors.

AREA OF RISK

Improper data management may result in valuable test resources being lost or invalid conclusions being reached.

A poorly written or executed data collection plan may result in the wrong data being collected, the data not being clearly marked, the loss of data due to improper storage and distribution, or the loss of data during the reduction process.

OUTLINE FOR REDUCING RISK

a.   **Data Collection Concept.**

Is the data collection concept for the systems to be tested rational and executable?

The data collection concept should provide pertinent information, including a list of key personnel, resources, and procedures. The concept should indicate the degree to which instrumented and manual data collection techniques will be used, with a discussion of why each technique is used. The concept should discuss the questionnaires and structured interview forms that will be used to support the qualitative measures. There should also be a discussion of how and when opinions, interviews, and observations of the players, controller, data collector, and test directorate personnel will be gathered.

> The data collection concept for a missile did not specify the forms or formats to be used for the collection of test player comments. During the field assembly of the missile, it was evident that troops could not assemble the missile with the equipment provided. The Test Director halted the test and had the troops interviewed to clearly identify the problem. After the interviews were conducted, it was determined that various interviewers had placed undue emphasis on specific aspects of the problem and had ignored other areas.

b.   **Data Collection and Processing System.**

Is there a collection and processing system in place and do the test personnel understand what they must do to use the system?

The collection and processing system should describe how the data are to be organized, reduced, verified, managed, controlled, and stored. The test plan for the tested system will list each data requirement and its means of collection. The SA should consult the test plan to determine if collection and processing systems are arranged according to plans and if there are data requirements that will not be collected because of the process. The SA also may determine whether the data are properly organized and stored, according to collection source. The data flow diagram (in the data collection plan) is a good source for information.

> During the test of a armored vehicle, the test team did not correctly execute the data collection plan. This resulted in a disorganized data collection activity with piles of data forms, questionnaires, and other documents not being properly recorded, filed, and stored. By the third week of testing, the Chief of Data Collection was unable to determine, from the documents, if trials were conducted at day or night, what the specific weather conditions were, or the degree of slope that the vehicle was operating on during the trial. As a result of this improper data management, an additional week of testing was required to fill in blanks in the data matrix.

c.   **Quality Control of Test Data.**

Are the test data being independently validated in a timely manner?

There is a need to quickly validate the test data to determine if additional test trials are required to answer operational test issues. The SA should determine if the data collections are adequate to support the OT&E report. He should ensure that there are checks and procedures in place at the test site to preclude or detect and correct errors made in data collection, data entry, and data reduction. The procedure should also outline the process for making required corrections or changes in test data and how the audit trail for those corrections will be maintained.

> During the pretest, it was determined that data collected procedures were not validated; it was debated whether the test should begin before corrections were made. The Test Director was unsure if critical data were being accurately collected during the pretest. However, because of the resources involved, it was decided to start the test and complete validation during the early stages. Critical data were lost from the trials performed in the early days of the test. The test was stopped after one week to fully install the data collection procedures.

d.   **Data Integrity.**

What are the established procedures for ensuring the maintenance of data integrity?

The data collection plan will discuss the process through which each set of collected data is to pass before reaching the storage medium which supports the Test Director and the evaluator. Data flow diagrams identify where data are combined with other data, and where they are processed, scored, reorganized, validated, or otherwise manipulated. The data collection plan should describe the data manipulation process at each step, along with rules and procedures for manipulation. The SA may wish to review the data collection plans and to include the flow diagrams to ensure that careful consideration has been given to data integrity.

> Control of the data was lost during the transfer from the test site, through a support contractor, to the test agency. It was determined that an audit trail was not established during the planning for data collection, transfer, and storage. This situation allowed data to be reorganized and manipulated without consideration of the effects of these activities on their evaluation. As a result, some data could not be used because their authenticity could not be assured.

6.8 TEST SCENARIOS

While visiting the test site, the Staff Assistant should examine and record the details of the test scenario that is being used. The comparison to the scenario that was described in the test plan is an important part of the on-site assessment. The test scenarios should provide realism, as well as an opportunity for test personnel to collect data on the system's effectiveness and suitability. Operational test scenarios normally are well planned, but due to location, instrumentation, and data collection requirements, they may not be well executed.

AREA OF RISK

Operational control often is lost and trials become force-on-force tactical maneuvers.

The Staff Assistant may determine if operational control of the tactical forces is maintained to ensure that test objectives are being met. Scenarios that are designed to support the evaluator's requirements must be closely followed to ensure that data collection is accomplished. Tester control, in a force-on-force operational test, is not always easy, because tactical units tend to perform in direct proportion to their level of training and motivation. During initial trials, troops will be highly motivated and tend toward being uncontrollable in their zeal to "win the war." However, as the test wears on, test personnel will become bored. Therefore, test scenarios have to be carefully monitored to ensure that troops react with enough "gusto" to provide realism.

OUTLINE FOR REDUCING RISK

a. **Operational Uncertainty.**

Does the test plan dictate that a level of operational uncertainty be maintained during the test?

To conduct realistic operational tests on some systems requires that actions by the threat systems be uncertain to the player personnel. This can mean uncertainty of time, location, type of action, direction, etc. The SA needs to assess the "reasonableness" of the threat scenarios.

> The enemy attack forces were scheduled to initiate a new test event each day after the forces reached the exercise area. As a result, the players had a rough idea of when the attack would begin. After three days, this scheduling became obvious and the Test Director took control of initiating the attack forces.

b. **Approved Doctrine and Tactics Used During the Test.**

Are the tactics and doctrine used by the test system and the threat systems approved by the responsible agency?

There usually is considerable disagreement about how the enemy will fight a new weapon system. For example, in a high intensity conflict, will the enemy-attack helicopter be deployed nap of the earth, low level, or at altitude? The doctrine and tactics for engagements of these helicopters will depend on an assessment of the most probable methods of deployment. The SA should determine if the Service has clearly established doctrine and tactics, based on the approved STAR, for the deployment of the tested system. He should determine if the test scenarios follow the approved doctrine and tactics.

> During the testing of an attack aircraft, threat crew members decided to use tactics that did not conform to approved doctrine. These tactics caused wide divergence in the planned scenarios and several days of testing were disrupted while the crews were properly briefed. Proper tactics and doctrine were included in the test events thereafter.

c. Tactical Operations Center Activities.

Is there a tactical operational center located in the field that is operated by trained tactical operations personnel?

To maintain control of tactical forces during a force-on-force operational test, it is important to have a fully manned tactical operations center located in the field with the operating forces. This center should be manned by fully trained operations personnel who are detailed from a tactical unit. The forces must be carefully briefed on the scenarios and the objectives of the operational test and evaluation. The SA should ensure that these personnel understand that good control of the maneuvering forces must be maintained in order for the test personnel to collect the data required to evaluate the system being tested.

> Prior to an OT event that included an armor engagement, the tanks being tested where driven for a number of hours to simulate a cross-country movement; as a result, the systems in the tanks were in a condition similar to what they might reflect in actual use. In this situation, to go directly from maintenance facilities to the engagement areas was not realistic.

d. Testing Terrain and Climatic Conditions.

Are environmental factors for the OT realistic for the system's mission profile?

Because of the availability of operational test sites, it is very difficult to find the desired terrain and climatic conditions in order for OT to be conducted under conditions that are fully representative of the system's intended deployed operational mission profile. Also, tradeoffs will have to be made because of safety and operational constraints. The SA should determine if the site selected for the test is representative of the terrain and climatic conditions required to provide realistic testing.

> During the Phase X OT of an Army vehicle system, heavy rains inundated a number of test site areas. A test event was added to demonstrate the vehicle recovery capability and the associated training, equipment, and procedures.

Chapter 7

OT&E REPORT

The OT&E report can be viewed as the most important document in the OT&E process. While it is true that the TEMP provides the foundation for all of the operational test and evaluation and the test plan provides the details for the information in the TEMP, the OT&E report is the product of all the test planning, management, and actual conduct of the T&E. The OT&E report is the document that provides the T&E results to the decisionmakers. It further serves to complete the process and realize the purpose of operational test and evaluation, and to provide information upon which to base decisions. Without clear and complete reporting in the test report, test planning and test conduct may have little impact on the acquisition program decision process.

There are a number of different OT&E reports; quick look reports, informal reports, briefing reports, and formal reports for different acquisition phases. Each of these OT&E reports must be considered for its individual purpose, and the phase the acquisition program is in. The report should show the audit trail through the program milestones and present the program's progress. It should summarize the planned testing, what happened during the test, and the test results.

In reviewing the OT&E report, the DOT&E Staff Assistant has two objectives. First, he should ensure that the report accurately reflects the test and evaluation that took place, and that it contains current, complete, and accurate data. Second, he should review the test and evaluation results on the system, including the issues and deficiencies, and recommend a DOT&E position to be reported to the appropriate decisionmaking forum.

There is no standard Department of Defense format for an OT&E report; each Service has developed its own approach to documenting the required information. Within the Army, there are two separate reports for OT&E activity. The test report (TR) is the primary record of the operational test and the findings and facts that resulted from the testing. The independent evaluation report (IER) documents the evaluation of the test data. These two documents together provide the composite information that is provided in the OT&E reports of the other Services.

From an operational suitability standpoint, the operational test and evaluation report is the concluding document in the process of determining if a system is suitable for operational use. The report should highlight the background of the test, i.e., the set of circumstances and decisions that determined what testing was conducted. Most operational testing has limitations, so it is important that the limitations that affect the suitability results are identified and documented in the OT&E report. The framework for reporting the testing results is the evaluation criteria that were developed prior to the actual testing. These criteria include the suitability critical issues and the associated characteristics, parameters, and thresholds. The test that was conducted should be summarized and any unexpected changes from the testing that was planned should be identified and documented. Finally, the report should discuss the test data and results, as well as provide an evaluation of the system and conclusions about the system's operational suitability.

Table 7-1 OT&E Report Formats

OPERATIONAL SUITABILITY GUIDE	ARMY TR (DAP 71-3)	ARMY IER (DAP 71-3)	AIR FORCE (AFOTECR 55-1)	NAVY (OTD GUIDE)	MARINE CORPS (MCOTEA)
7.1 Background Issues	1.0 Introduction	1.0 General	I. Purpose and Background	1. Purpose 2. Equipment Description 3. Background 4. Scope 7. Critical Operational Issues	1. Introduction 2. System Description 3. Objectives and Issues
7.2 Limitations					
7.3 Evaluation Criteria					
7.4 Summary of Test Conduct	2.0 Test Results		II. OT&E Description III. Operational Effectiveness and Suitability	5. Project Operations 6. Results	4. Evaluation
7.5 Test Results					
7.6 Conclusions	* Conclusions Are Provided in Some Reports	2.0 Findings of Evaluation	V. Summary of Conclusions and Recommendations	9. Conclusions 10. Recommendations	
	Appendices	Appendices	IV. Service Reports	8. Operational Considerations	

The templates that follow address each of the major content areas included within OT&E reports. These areas are:

    7.1   Background Issues
    7.2   Limitations
    7.3   Evaluation Criteria
    7.4   Summary of Test Conduct
    7.5   Test Results
    7.6   Conclusions

Table 7-1 was prepared from the Services' policies on OT&E. It shows how the structure of this book relates to the major sections of the respective Services' OT&E reports.

7.1 BACKGROUND ISSUES

The background issues include those major events and information that are needed to place the OT&E results into proper perspective. Items in this category include program direction on the conduct of the OT&E, and the results and limitations of earlier operational testing as well as significant operational issues.

AREA OF RISK

Unclear statements of issues may result in an incomplete or biased view of what the test results mean.

If the background is not summarized adequately, the testing and T&E results may not be placed in the proper perspective. The T&E activity on a program is a continuum of activity that includes periodic reports. To gain maximum benefit from these reports, the context of the testing, the meaning of the evaluation, and full consideration of the results is essential. If the purpose of the testing and the background are not described adequately, the reader may formulate an incomplete or biased view of what the test results mean.

OUTLINE FOR REDUCING RISK

a.   **Previous Test Phases.**

Are the important operational suitability aspects of the previous test phases summarized?

The previous test phases provide the fundamental starting point for the test that is being reported on. The summary of these earlier test periods should include the important areas that were satisfactorily tested and also summarize the significant results. It should include which operational suitability areas were tested well and had good results, which areas were not included in the test, and which areas were tested but did not meet their requirements.

> The previous system tests were summarized as they apply to the major RAM characteristics. The reliability chart showed the stated quantitative requirement (in mean time between operational mission failure - MTBOMF) for each of the three configurations. The test results for each of the completed OT phase were listed. For reliability, the point estimate value and the value at the 80 percent lower confidence limit were shown in the chart.

b.   Previous Management Decisions.

Did previous management decisions drive the operational suitability test planning?

Are the decisions that directed the test and the issues involved in those decisions described in enough detail to permit the reader to understand the context of the decision? Management direction that resulted from previous testing may highlight critical or risk areas; an understanding of these concerns can assist in placing the operational suitability results into proper perspective.

> Management review of the previous testing results focused attention on the progress in the diagnostics area. The maturity level of the built-in test during the previous test phase was less that expected. This area was highlighted by the Acquisition Decision Memorandum as an area of risk that should received additional attention during the next OT&E phase.

c.   Operational Suitability Characteristics.

Are suitability characteristics included in the system summary?

The system summary within the test report should be a relatively short discussion of what was tested, emphasizing the mission or function of the system. Any system attributes, i.e., new technology or capability, high risk areas, etc., that are the reasons for critical operational issues (COIs) should definitely be included. The support concept also should be summarized, since this is needed for assessing how complete the operational suitability portion of the OT&E was.

> The system summary includes an outline of the maintenance concept planned for the system. The summary outlines how the system will be supported at each level of maintenance and what maintenance will be performed at each level.

d.   System Differences.

How was the tested system different from the planned operational system? What are the resulting implications for suitability testing?

The systems under test, particularly in the early stages of OT&E, may be significantly different from the planned operational system. These early systems generally will not be from the production line that is planned for the full-rate production and, therefore, will not have the full benefit of stable production processes. The system differences should be discussed as they relate to limitations on the suitability testing and the need for any additional suitability testing in a later phase.

> The test report indicated that the operational software that was available during the test was not the version planned for the production systems. Because of this difference, there were some diagnostic and maintenance operational capabilities that were not verifiable during the operational suitability testing.

7.2 LIMITATIONS

All major tests have limitations. These may include limits in the operating environment, the length of the test period, or the type or number of test units, personnel, or supporting devices. The list of limitations highlights the potential risks involved in assuming that test results are totally indicative of what might be expected in the operating environment.

AREA OF RISK

Insufficient identification of limitations may result in inaccurate assessment of the OT&E results.

Limitations that occur during the actual testing may not be highlighted in the test report. Such limitations could result from the lack of realism of the test environment, the test duration, the number of test articles, or the availability of test hours. The test articles might not be of the latest configuration or capability. Without visibility of the limitations, suitability implications may be overlooked. The risk in this area is that limitations are not identified sufficiently for the decisionmaker to know how to assess the OT&E results.

OUTLINE FOR REDUCING RISK

a.   **Operational Suitability Test Limitations.**

Was there a summary of all known test limitations that would affect suitability?

Operational suitability limitations may include many areas. Limited number of test hours can be a limitation on evaluating the level of reliability and maintainability. Others may be the result of the logistics support during the test period being not representative of the planned operational support. Test equipment may not be available. All operational suitability limitations should be summarized to permit the reader to place the test results into perspective.

> There was an inadequate number of test hours on selected host vehicles. A complete reliability evaluation of the system is dependent on sufficient test hours to provide statistical confidence in the results. Lack of test hours and a compressed test period will reduce the statistical confidence in the OT&E results.

b.   Additional Limitations During Testing.

Were there any limitations that developed during the conduct of the actual testing? Are they presented and their impact on operational suitability discussed?

Additional limitations beyond those in the test plan may surface once the test has started. Specific test articles may not be available as planned. Other program priorities may dictate a revision to the activity that was planned in the OT&E test plan. The test equipment planned for the test site may not be available. Other items, such as technical documentation, may be delayed and not included in the operational suitability evaluation.

> The software for testing the radar system components on the XX-537 test station was not available at the test site as planned. As a result, the second-level maintenance for the radar was not included in the test.

c.   Environmental Differences.

Were there significant differences between the test environment and the expected operational environment?

Test limitations could result for significant environmental differences. These differences could be major factors in assessing the acceptability of the operational suitability results. The test report should identify the major conditions that were different. This could include the ratio of support assets to test articles, the skill levels of the maintenance personnel, or the depth of training that the maintenance personnel received.

> The maintenance team that supported the system during the testing was generally of a higher skill level than the planned maintenance organization. Analysis of specific task times was performed to determine the effect of this skill difference and the resulting impact on the measured mean time to repair.

d.   Suitability Elements Not in the Test.

What suitability elements were not able to be evaluated during the test?

In most OT&E phases it is common for some support elements to not be present during the testing. An example is second-level test equipment that is not yet developed to the stage where it is ready for the OT&E environment. Other elements may have items substituted that are significantly different, such as factory test equipment used instead of the planned operational test equipment. The effect of these situations must be considered to identify the areas of suitability that are not yet evaluated. These risk areas should be considered by the decisionmakers and included in future phases of test and evaluation.

> For the test period being reported, the second level of maintenance was performed by the hardware contractor at his facility and therefore was not part of the evaluation.

7.3 EVALUATION CRITERIA

The evaluation criteria should be derived from the requirements of the user organizations. There also may be qualitative criteria for some of the operational suitability elements.

AREA OF RISK

If the reported results are not related to the user's stated requirements, then the test and evaluation report may not be meaningful.

The proper evaluation of suitability test or analysis results is dependent on the pre-established criteria. These criteria must be related to the user's stated requirements. The test and evaluation report may not be meaningful if the results are not related to the user's stated requirements.

OUTLINE FOR REDUCING RISK

a.   **Criteria.**

Were there established criteria for each of the operational suitability characteristics?

To be meaningful, each operational suitability characteristic should have evaluation criteria drawn from the user's stated requirements and stated in the OT&E test plan. Was enough detail provided to define the criteria? For example, is there some indication of how failures are defined? Was a reference listed that included the failure definition?

> The reliability criterion is stated in terms of mean miles between unscheduled maintenance actions (MMBUMA). The system shall achieve 200 MMBUMA.

b.   **Source of the Operational Suitability Criteria.**

What is the source of the quantitative and qualitative criteria for the operational suitability elements?

When the criteria in the operational suitability are stated in the test report, they give the reader a framework for evaluation of the test results. To provide a complete picture, the test report should identify what the sources were for the evaluation criteria. This is particularly important in the statement of any qualitative criteria. Most quantitative criteria can be traced to requirements documents, but this is not always true of the qualitative criteria. (OPTEVFOR does not use criteria on the qualitative measures.)

> The system reliability, maintainability, and availability requirements were specified in the July 1987 revision of the system Required Operational Capability (ROC).

c.   Quantitative Reliability, Availability, and Maintainability (RAM) Measures.

Are the suitability characteristics for RAM stated in quantitative terms?

For most systems, operational suitability characteristics such as reliability, availability, and maintainability (RAM) can be expressed with quantitative parameters. The evaluation criteria for these areas therefore should be stated quantitatively and used in evaluating the test results. The definition of the appropriate terminology is an essential part of understanding these numerical values.

> The fixed-installation communication set was required to have a mean time between operational mission failure (MTBOMF) of greater than 500 hours.

d.   Confidence Limits.

Are there statements of confidence limits for the RAM quantitative criteria?

The measurement of statistical parameters, such as reliability, can require considerable test time. If limited test time is available, as is often the case, then the measured value is expressed with a degree of statistical confidence. Background information on determining confidence levels from test data is discussed in DoD 3235.1-H, "Test and Evaluation of System Reliability, Availability, and Maintainability--A Primer." This document discusses the mathematics of test statistics, but does not aid in deciding what level of confidence is needed. (While OPTEVFOR uses confidence calculations in test design, these factors usually are not mentioned in their test reports. The other OTAs use confidence levels when they are considered appropriate.)

> The mission reliability of the system shall be measured in terms of the mean time between operational mission failures (MTBOMF). The point estimate MTBOMF that results from the scoring conference data shall be used to compute an 80 percent lower one-sided confidence interval. The 80 percent confidence value shall exceed the requirement stated in the approved ROC.

7.4 SUMMARY OF TEST CONDUCT

The OT&E report must summarize the actual suitability testing that was performed. This summary must include an adequate description of the operational suitability testing and present all significant changes from information that was presented in the OT&E plan.

AREA OF RISK

Omission of critical information may alter the meaning of the test results.

A change in the planned suitability test activity may result in significant changes to the meaning of the test results. The test report should present the description of what was actually done and what the results were. Some of the information on deficiencies may be omitted from the summary in an effort to condense the information or to protect the results from further scrutiny. The DOT&E Staff Assistant must have enough visibility into the actual testing to know if the summary is complete and accurately reflects the suitability T&E that was performed. The risk in the summary is that some critical information will not be included.

OUTLINE FOR REDUCING RISK

a.   **Comparison to Test Plan.**

Was the operational suitability test conducted consistent with the suitability test planning?

The discussion should highlight any differences between the suitability test as planned and as conducted. Any differences most likely are at the detailed level; thus, some detailed discussion is required to point out the differences, as well as any impact these differences may have on the test results.

When operational testing is performed, changes often are required because of conditions that were not foreseen when the OT&E plan was prepared. These changes must be summarized in the OT&E report. They should be discussed in enough detail to permit the reader to assess what impact these items had on the test and the test results.

> During test phase II, the XYZ test set was not available as had been planned. As a result, the reliability of the test set, the compatibility of the test set with the system, and the testability of the system were not evaluated as had been planned during test phase II.

b.   Unexpected Limitations.

Were there any unexpected limitations to the operational suitability portion of the test?

Unexpected test limitations may result from the differences discussed above; but they may result also from unexpected suitability-related factors, such as weather and support personnel.

> The results of the operational testing showed that the maintenance training that was conducted prior to the start of the test was inadequate in the area of radar fault isolation and troubleshooting.   This deficiency precluded a complete evaluation of maintainability, maintenance documentation, and the diagnostics system for the radar.

c.   Summary of the Test Performed.

Is the summary of the test that was performed complete and unambiguous?

Based upon "first-hand" knowledge of the test conduct, the Staff Assistant should assess the summary in the OT&E report.   Different authors will describe suitability-related events differently. Identical summary descriptions are unlikely, but the objective is to have a discussion that includes a fair and accurate summary of the significant aspects of the suitability testing that was conducted. The facts concerning what was done and what occurred during the suitability testing should be clearly stated. The activities should be related to the test planning or other reference documents.

> The search radar system was flown on 43 sorties. The Type I mission profile was flown as described in the system OT&E test plan. The total flying hours on the system were 168 hours. The system experienced three mission failures during the test period. The point estimate MTBOMF was 56 hours.

7.5 TEST RESULTS

The summary of the test results is the major section of the OT&E test report. The information here should support the suitability conclusions reached and provide a basis for the readers to form judgments that agree with the major conclusions. The level of detail that is provided should permit the readers to integrate independent thoughts with the details of the suitability test results.

AREA OF RISK

Poor presentation of important suitability factors may lead to incorrect conclusions.

The test results must be discussed in enough detail to support any conclusions that the system meets or does not meet its suitability criteria for the phase being assessed. If adequate detail is not presented, the major conclusions may not be accepted and may not be supported by the reader. The major findings should be highlighted in a manner that gives the reader insight into the important suitability results that evolved from the test and the evaluation of the test data.

OUTLINE FOR REDUCING RISK

a.   **Presentation of Suitability Results.**

Are the major suitability findings presented in an understandable way?

The major operational suitability findings should be in areas related to the suitability critical operational issues (COIs). There may be important operational suitability findings in other areas as well. The DOT&E Staff Assistant needs to review the findings and ensure that all important areas are addressed.

> Overall Evaluation: The operational suitability of the system is satisfactory. Reliability and maintainability exhibited in OT&E exceeded the stated requirements.

b.   Projection Methods.

Was the method used to project RAM values from the test results to mature values examined and validated?

If the test report states values for the mature projection of reliability and maintainability (R&M), is the method of projection described in enough detail to evaluate its applicability? Some operational testing agencies never use projections, others use them often. There are several projection methods available for use. When they are used, the validity of the methods and applicability of the method to the situation at hand must be assessed. The use of a questionable projection method may cause the decisionmaker to place little value on the OT&E results. A projection should never be reported as a test result--the test result should be an observed value.

> The reliability test results were projected to the mature system values using methods from Military Handbook 189. The applicability of the method to the system was validated by using growth experience with similar systems, and by reviewing the reliability growth plans presented by the program manager.

c.   Confidence Levels.

Are confidence levels stated for the quantitative R&M values that resulted from the tests?

The use of limited test data to provide measures of reliability and maintainability always includes room for statistical error in the quantitative estimate. The likelihood of the error is shown by the statistical confidence associated with the test measurement. For each appropriate measurement, a statistical confidence should be stated. (While OPTEVFOR uses confidence calculations in both test design and in evaluation, the confidence values are not included in OPTEVFOR test reports.)

> Reliability: An 80 percent lower one-sided confidence interval was calculated for the point estimate value of the MTBOMF. Operational mission failures were determined by a formal scoring conference in accordance with the approved failure definition and scoring criteria. The 80 percent confidence value exceeded the requirement stated in the approved ROC.

d.   Impact of Suitability Results.

Is the impact or consequence of the major suitability findings stated?

When operational suitability results are compared with their criteria, the report should state that for this particular suitability area, the system is acceptable, marginal, or unacceptable, or offer some other judgment. The impact or consequence of these conditions on the system should be included in the test report. The DOT&E staff assistant should have an understanding of the situation and its impact on the system's operational capability.

> The stability of the ammunition trailer is marginal. The implication of this deficiency is that the trailer will be unable to carry a full load of ammunition and travel at normal speeds on an unpaved road without tipping over.

7.6 CONCLUSIONS

The conclusions that are reached as a result of the operational testing should be clearly stated in the OT&E report. In this section, there should be a clear statement of whether the system was considered operationally suitable.

AREA OF RISK

Poorly stated or omitted conclusions can result in erroneous management decisions.

Not all test reports contain conclusions on the acceptability of the system. Conclusions should be stated that indicate the major points that are drawn from the suitability test results and the evaluation of the data against test criteria. Was the test passed and what did the test prove? Which areas are acceptable and which are not? What is the meaning of the deficient areas relative to the progress of the system toward its suitability for operational use?

OUTLINE FOR REDUCING RISK

a. **Operational Suitability Conclusion.**

Does the report state whether the system is considered operationally suitable? Is this operational suitability assessment a composite view of the suitability elements?

How does the report deal with a situation in which some of the elements are satisfactory and some are not? Are the unsatisfactory items highlighted? Those items that are not within the user's stated needs or that do not meet the OT&E criteria must be assessed for impact on the system's overall suitability for fielding. A single conclusion in the operational suitability area may be difficult. This requires the combination of the many suitability elements--some of which may be good, some marginal, and some not evaluated--into a single judgment. The suitability test results must be evaluated within the context of the planned operational use of the system and a judgment must be made as to whether the system meets the user's needs. Suitability COIs are used to focus attention on areas that should receive increased weighting in any total assessment.

> Although the system meets its operational mission performance and the RAM requirements, it is not suitable for fielding without major changes to the planned logistic support. The system is judged as not operationally suitable because of deficiencies in second-level support equipment and the quantity of the planned spares.

b.   Operational Effects of Any Adverse Test Results.

What is the operational implication of the operational suitability deficiencies?

When an operational suitability deficiency is highlighted in the test report conclusions, the report should include the implication of this deficiency on the operational use of the system. How is the system limited in its use? What missions or uses are degraded by the deficiency? What additional resources are required to compensate for the existence of the deficiency?

The automatic diagnostics system for the weapon system was deficient in the area of fault isolation. The evaluation criterion for fault isolation was:

Isolate faults to one SRU -- 90 percent; two SRUs -- 95 percent.

Faults were isolated to one SRU -- 72 percent; two SRUs -- 84 percent.

Additional unit-level spares will be required to support the initial operating units until the diagnostics deficiencies are corrected.

Chapter 8

SERVICE OTAs COMMON RELIABILITY, AVAILABILITY, AND MAINTAINABILITY (RAM) TERMINOLOGY

The Military Services Operational Test and Evaluation Agencies (OTAs) have agreed to common methods to be used during testing that involve more that one Service. This is defined as Multi-Service OT&E. In 1989, the four Service OTAs agreed to a listing of "Common Reliability, Availability, and Maintainability Terms for use in Multi-Service OT&E Test Programs." This listing is contained in Annex A to the OTAs Memorandum of Agreement and is included in this book for reference.

The purpose of the listing is to provide a common set of terminology that can be used during testing of systems being acquired as "Multi-Service" programs. A Multi-Service program is one that is acquiring the same or similar systems for use by more than one of the Military Services. This could, for example, mean an electronics system or missile that is being developed for use by the Air Force and the Navy, or a radio or navigation system that is planned for use by the Army, Navy, and Marine Corps.

The testing of these Multi-Service programs is a complex undertaking. Not only may there be two (or three) operating environments, but the operating missions, scenarios, and measures of mission success may be different as well. In the suitability area, one of the complexities is that the Military Services have different means of measuring the quantitative measures of suitability. These are usually associated with reliability, availability, and maintainability (RAM). With the differences of measures, scenarios, etc., it is possible for a Multi-Service system to achieve its suitability requirements as defined by one Service, and to be deficient using the criteria of the second Service. The OT might have major limitations from the view of one Service, but be acceptable in the eyes of another Service.

The goal of the common RAM terms was to remove one of the potential problems, that of having multiple terms for the same parameter, or having different definitions for the same term. The Annex to the Service Memorandum of Agreement includes common terms and also includes listing and definitions for terms that are used by each of the Service OTAs for measuring suitability of the systems being operationally tested. Table A-1, Annex A, provides a summary comparison of the Multi-Service and Service-Unique terms and should be reviewed for an overview of the material in the Annex.

ANNEX A

COMMON RELIABILITY, AVAILABILITY, AND
MAINTAINABILITY (RAM) TERMINOLOGY

1. <u>Purpose</u>. This Annex provides the policy and common RAM terminology for the quantitative portion of MOT&E suitability evaluations.

2. <u>Background</u>. MOT&E common terms are intended to convey the same meaning to all Services. Therefore, they avoid terms used elsewhere with different meanings. Existing terms used by one or more Services were selected when possible. Table A-1 compares the RAM terms to be used for multiservice testing, as described in this Annex, with the relative service-unique RAM terms currently in use. Other relevant, service-unique RAM terms are provided in appendices to this Annex.

3. <u>Policy</u>

    a. Common terms described in this Annex will be used as appropriate in multiservice test reports. If additional terms are necessary, they should be agreed upon and clearly defined by all participating agencies.

    b. Multiservice terms selected will be included in the multiservice TEMP.

4. <u>Common RAM Terms for MOT&E</u>

    a. <u>Reliability</u>. Reliability consists of two major areas: mission reliability and logistics support frequency.

        (1) Mission reliability is the probability a system can complete its required operational mission without an operational mission failure (OMF). An OMF is a failure that precludes successful completion of a mission, and must be specifically defined for each system. For some systems, mission reliability may be better expressed as Mean Time (miles, rounds, etc.) Between Operational Mission Failure (MTBOMF). (See paragraph 5 for definition.)

        (2) Logistics support frequency is a representative time between incidents requiring unscheduled maintenance, unscheduled removals, and unscheduled demands for spare parts, whether or not mission capability is affected. Logistics support frequency may be expressed as Mean Time Between Unscheduled Maintenance (MTBUM). (See paragraph 5 for definition.)

149

b. <u>Maintainability</u>. Maintainability consists of three major areas: OMF repair time, corrective maintenance time, and maintenance ratio. Maintainability may be expressed as (1) Mean Operational Mission Failure Repair Time (MOMFRT), (2) Mean Corrective Maintenance Time (MCMT), (3) Maximum Time To Repair (MaxTTR), and (4) various maintenance ratios, e.g., Maintenance Man-Hours Per Operating Hour, Mile, Round, etc. (See paragraph 5 for definitions.)

c. <u>Availability</u>. Operational availability (A_o) is the probability that a system will be ready for operational use when required. (See paragraph 5 for definition.)

d. <u>Diagnostics</u>. Diagnostics addresses the ability of integrated diagnostics (automated, semi-automated, and manual techniques taken as a whole) to fault-detect and fault-isolate in a timely manner. Diagnostics may be expressed as (1) the percent of correct detection given that a fault has occurred (P_{cd}), and (2) Mean Time To Fault Locate (MTTFL). (See paragraph 5 for definitions.)

5. <u>Definitions for Multiservice Terms</u>

a. <u>Mean Time Between Operational Mission Failures (MTBOMF)</u>: The total operating time (e.g., driving time, flying time, or system-on time) divided by the total number of OMFs.

b. <u>Mean Time Between Unscheduled Maintenance (MTBUM)</u>: The total operating time divided by the total number of incidents requiring unscheduled maintenance.

c. <u>Mean Operational Mission Failure Repair Time (MOMFRT)</u>: The total number of clock-hours of corrective, on-system, active repair time, which is used to restore failed systems to mission-capable status after an operational mission failure (OMF) occurs, divided by the total number of OMFs.

d. <u>Mean Corrective Maintenance Time (MCMT)</u>: The total number of clock-hours of corrective, on-system, active repair time due to all corrective maintenance divided by the total number of incidents requiring corrective maintenance.

e. <u>Maximum Time To Repair (MaxTTR)</u>: That time below which a specified percentage of all corrective maintenance tasks must be completed.

f. <u>Maintenance Man-Hours Per Operating Hour (MMH/OH)</u>: The cumulative number of maintenance man-hours during a given period divided by the cumulative number of operating hours. If appropriate, other terms such as miles or rounds may be sub-

stituted for hours. Scheduled as well as corrective maintenance, in keeping with the users maintenance requirements,
are included without regard to their effect on mission or
availability of the system.

   g. Operational Availability (Ao): A_0 is either the total
uptime divided by the uptime plus downtime when operated in an
operational mission scenario, or the number of systems that are
ready divided by the number possessed.

   h. Percent of Correct Detection Given That a Fault Exists
(Pcd): The number of correct detections divided by the total
number of confirmed faults.

   i. Mean Time To Fault Locate (MTTFL): The total amount of
time required to locate faults divided by the total number of
faults.

APPENDICES:

1 - Army Terms and Definitions
2 - Navy Terms and Definitions
3 - Marine Corps Terms and Definitions
4 - Air Force Terms and Definitions

TABLE A-1. COMPARISON OF MULTISERVICE AND SERVICE-UNIQUE TERMS

CATEGORY	MULTI-SERVICE	ARMY	NAVY	AIR FORCE	MARINES
RELIABILITY	MTBOMF MTBUM	MTBOMF MTBUMA	MTBMCF MTBF	MTBCF MTBME	MTBOMF MTBUMA
MAINTAINABILITY	MOMFRT MCMT MaxTTR MMH/OH	NONE MTTR MaxTTR MR	MTTR NONE NONE DMMH/FH	NONE MRT MaxTTR MMH/OH	NONE MTTR MaxTTR MR
AVAILABILITY	A_o	A_o	A_o	A_o AND OTHERS	A_o
DIAGNOSTICS	P_{cd} MTTFL	P_{cd} AND OTHERS	P_{cd} AND OTHERS	P_{cd} AND OTHERS	P_{cd} AND OTHERS

ARMY TERMS AND DEFINITIONS

1. Purpose. This Appendix provides the RAM terms and definitions most relevant to this Annex and used within the Army in conducting and reporting OT&E activity in accordance with AR 702-3. They are included in the Memorandum of Agreement so as to assist other Services in understanding RAM terms as used by the Army. Terms used by other Services are included in Appendices 2, 3, and 4.

2. Definitions

   a. Durability: A special case of reliability; the probability that an item will successfully survive to its projected life, overhaul point, or rebuild point (whichever is the more appropriate durability measure for the item) without a durability failure. (See Durability Failure.)

   b. Failure: The event, or inoperable state, in which an item or part of an item does not, or would not, perform as previously specified. (See MIL-STD 721.)

   c. Failure, Critical: A failure (or combination of failures) that prevents an item from performing a specified mission. (Note: Normally only one failure may be charged against one mission. Critical failure is related to evaluation of mission success.)

   d. Failure, Durability: A malfunction that precludes further operation of the item, and is great enough in cost, safety, or time to restore, that the item must be replaced or rebuilt.

   e. Failure Mode: The mechanism through which failure occurs in a specified component (for example, short, open, fatigue, fracture, or excessive wear). (See MIL-STD 721.)

   f. Inherent RAM Value: Any measure of RAM that includes only the effects of an item design and its application, and assumes an ideal operating and support environment.

   g. Maintainability: A measure of the ability of an item to be retained in, or restored to, a specified condition when maintenance is performed by personnel having specified skill levels using prescribed procedures and resources.

   h. Maintenance Ratio (MR): A measure of the maintenance manpower required to maintain a system in an operational environment. It is expressed as the cumulative number of direct

maintenance man-hours (see AR 570-2) during a given period,
divided by the cumulative number of system life units (such as
hours, rounds, or miles) during the same period. The MR is
expressed for each level of maintenance and summarized for
combined levels and maintenance. All maintenance actions are
considered (that is, scheduled as well as corrective, and
without regard to their effect on mission or availability of
system). Man-hours for off-system repair of replaced
components are included in the MR for the respective level.

    i. Maximum Time To Repair (MaxTTR): That time below
which a specified percentage of all corrective maintenance
tasks must be completed. When stated as a requirement, the
MaxTTR should be stated for organizational and direct support
levels of maintenance. MaxTTR is used as an "on-system"
maintainability parameter; it is not used for the off-system
repair of replaced components.

    j. Mean Time Between Essential Maintenance Actions
(MTBEMA): For a particular measurement interval, the total
number of system life units divided by the total number of
nondeferrable maintenance actions. This parameter indicates
the frequency of demand for essential maintenance support and
includes incidents caused by accidents, maintenance errors, and
item abuse. (Not included are crew maintenance completed
within a specified number of minutes, maintenance deferrable to
the next scheduled maintenance, system modification, and test-
peculiar maintenance.)

    k. Mean Time Between Operational Mission Failure (MTBOMF):
A measure of operational effectiveness that considers the
inability to perform one or more mission-essential functions.

    l. Mean Time Between Unscheduled Maintenance Actions:
Computed by the following formula:

$$\text{MTBUMA} = \frac{\text{Operating time}}{\text{Total number of unscheduled maintenance actions}}$$

    m. Mean Time To Repair (MTTR): The sum of corrective
maintenance times divided by the total number of corrective
maintenance actions during a given period of time under stated
conditions. MTTR may be used to quantify the systems maintainability characteristic. MTTR applies to the system-level
configuration; it will be used as an "on-system" maintainability
index and not for the repair of components. MTTRs will be
stated for the unit and the intermediate direct support levels
of maintenance along with the percentage of actions repaired at
each level.

n. **Mission Reliability (Rm)**: A measure of operational effectiveness. It is stated in terms of a probability of completing a specified mission profile or the mean time (or distance or rounds) between critical failures.

o. **Mission–Essential Functions**: The minimum operational tasks that the system must be capable of performing to accomplish its mission profiles.

p. **Off–System Maintenance**: Maintenance associated with the diagnosis and repair of components for return to stock.

q. **On–System Maintenance**: Maintenance necessary to keep a system in, or return a system to, an operating status.

r. **Operational Availability**: The proportion of time a system is either operating, or is capable of operating, when used in a specific manner in a typical maintenance and supply environment. All calendar time when operating in accordance with wartime operational mode summary/mission profile (OMS/MP) is considered. The formula is as follows:

$$A_O = \frac{OT + ST}{OT + ST + TCM + TPM + TALDT}$$

$$= \frac{\text{Total calendar time minus total downtime}}{\text{Total calendar time}}$$

Where: OT = The operating time during OMS/MP

ST = Standby time (not operating, but assumed operable) during OMS/MP

TCM = The total corrective maintenance downtime in clock-hours during OMS/MP

TPM = The total preventive maintenance downtime in clock-hours during OMS/MP

TALDT = Total administrative and logistics downtime (caused by OMFs) spent waiting for parts, maintenance personnel, or transportation during OMS/MP

Other forms of this equation are substituted depending on the system type (see AMC/TRADOC PAM 70-11) such as the inclusion of relocation time.

s. **Operational Mission Failure (OMF)**: Any incident or malfunction of the system that causes (or could cause) the

A-1-3

inability to perform one or more designated mission-essential functions.

t. Operational RAM Value: Any measure of RAM that includes the combined effects of item design, quality, installation, environment, operation, maintenance, and repair. (This measure encompasses hardware, software, crew, maintenance personnel, equipment publications, tools, TMDE, and the natural, operating, and support environments.

u. Reliability: The probability that an item can perform its intended functions for a specified time interval under stated conditions.

v. Reliability After Storage: This may be a stated requirement. If appropriate, it specifies the amount of deterioration acceptable during storage. Length of storage, storage environment, and surveillance constraints are identified. This requirement may not be testable; it may rely on an engineering analysis for its assessment before deployment.

w. System Readiness Objective (SRO): One of a group of measures relating to the effectiveness of an operational unit to meet peacetime deployability and wartime mission require-ments considering the unit set of equipages and the potential support assets and resources available to influence the units operational readiness and sustainability. Peacetime and wartime SROs will differ due to usage rate, OMS/MPs, and operational environments. Examples of SROs include: operational availability at peacetime usage rates, operational availability at wartime usage rates, sortie generations per given time frame (aircraft), and maximum ALDT (intermittent mission). SROs must relate quantitatively to system design parameters (RAM) and relate to support resource requirements. (See AR 700-127.)

A-1-4

APPENDIX 2

NAVY TERMS AND DEFINITIONS

1. Purpose. This Appendix provides the RAM terms and
definitions most relevant to this Annex and used within the
Navy in conducting and reporting OT&E activity in accordance
with COMOPTEVFORINST 3960.1E. They are included in the
Memorandum of Agreement so as to assist other services in
understanding RAM terms as used by the Navy. Terms used by
other services are included in Appendices 1, 3, and 4.

2. Definitions

a. Availability: A measure of the degree to which an item
is in an operable and committable state at the start of a
mission when the mission is called for at an unknown (random)
time. In OT&E, operational availability (A_o) is the usual
measure. (See Operational Availability.)

b. Failure, Critical: One that prevents the system from
performing its mission or results in the loss of some
significant mission capability.

c. Failure, Minor: One that affects system performance
but does not impact the ability to perform the mission.

d. Maintainability: The capability of an item to be
retained in or restored to specified conditions when
maintenance is performed by personnel having specified skill
levels, using prescribed procedures and resources, at each
prescribed level of maintenance and repair. MTFL, MTTR, and
MSI are frequently calculated in maintainability evaluations.

e. Maintenance Support Index (MSI): The ratio of total
man-hours required for maintenance (preventive plus corrective)
to the total operating (up) time. Frequently computed as part
of Test S-2 Maintainability.

f. Mean Flight Hours Between Failure/Mean Time Between
Failure (MFHBF/MTBF): See Reliability.

g. Mean Time To Fault-Locate (MTFL): The total fault-
location time divided by the number of critical/major failures.
Frequently computed as part of Test S-2 Maintainability.

A-2-1

h. <u>Mean Time To Repair (MTTR)</u>: Normally computed as part of maintainability, MTTR is the average time required to perform active corrective maintenance. Corrective maintenance is the time during which one or more personnel are repairing a critical or major failure and includes: preparation, fault location, part procurement from local (on-board) sources, fault correction, adjustment/calibration, and follow-up checkout times. It excludes off-board logistic delay time.

i. <u>Mission Reliability</u>: See Reliability.

j. <u>Operational Availability</u>: (See Availability for basic definition.) Operational availability is computed and reported as follows:

(1) <u>Continuous-Use Systems</u>: Operational availability shall be designated A_o and shall be determined as the ratio of system "uptime" to system "uptime plus downtime."

(2) <u>"On-Demand" Systems</u>: Operational availability shall be designated A_{od} and shall be determined as the ratio of the "number of times the system was available to perform as required to the total number of times its performance was required." (Note: "Total number of times its performance was required" shall be the number of times attempted and the number of times it was operationally demanded but not attempted because the system was known to be inoperable.)

(3) <u>Impulse Systems</u>: Operational availability shall be designated A_{oi}, and since A_{oi} and R are inseparable during testing, only reliability (R) shall be reported.

k. <u>Operational Effectiveness</u>: The capability of a system to perform its intended function effectively over the expected range of operational circumstances, in the expected environment, and in the face of the expected threat, including counter-measures where appropriate.

l. <u>Operational Suitability</u>: The capability of the system, when operated and maintained by typical fleet personnel in the expected numbers and of the expected experience level, to be reliable, maintainable, operationally available, logistically supportable when deployed, compatible, interoperable, and safe.

m. <u>Reliability</u>: The duration or probability of failure-free performance under stated conditions. In OT&E, reliability is usually reported in one of two ways:

(1) <u>Mean Time Between Failure (MTBF)</u>: For more-or-less continuously operated equipment, the ratio of total operating time to the sum of critical and major failures. MTBF is

A-2-2

sometimes modified to mean flight hours between failures (MFHBF).

(2) <u>Mission Reliability</u>: For equipment operated only during a relatively short-duration mission (as opposed to equipment operated more-or-less continuously), the probability of completing the mission without critical or major failure. Frequently expressed as exp (-t/MTBF), where "t" is mission duration and MTBF is as defined above.

A-2-3

APPENDIX 3

MARINE CORPS TERMS AND DEFINITIONS

1. <u>Purpose</u>. This Appendix provides the RAM terms and definitions most relevant to this Annex and used within the Marine Corps in conducting and reporting OT&E activity in accordance with FMFM 4-1 (Combat Service Support), TRADOC/AMC Pamphlet 70-11 (RAM Rationale Report Handbook), and DoD 3235.1-H (Test and Evaluation of System Reliability, Availability, and Maintainability, a Primer). They are included in the Memorandum of Agreement so as to assist other services in understanding RAM terms as used by the Marine Corps. Terms used by other services are included in Appendices 1, 2, and 4.

2. <u>Definitions</u>

    a. <u>Achieved Availability (Aa)</u>: Computed with the following formula:

$$A_a = \frac{OT}{OT + TCM + TPM}$$

Where: OT = Operating time
          TCM = Total corrective maintenance
          TPM = Total preventive maintenance

    b. <u>Administrative and Logistics Downtime (ALDT)</u>: The period of time that includes (but is not limited to) time waiting for parts, processing records, and transporting equipment and/or maintenance personnel between the using unit and repair facility.

    c. <u>Corrective Maintenance (CM)</u>: Maintenance that is performed on a nonscheduled basis to restore equipment to satisfactory condition by correcting a malfunction (unscheduled maintenance). The measure is Total Corrective Maintenance (TCM) time.

    d. <u>Depot Level Maintenance</u>: Maintenance that is performed by designated industrial-type activities using production-line techniques, programs, and schedules. The principal function is to overhaul or completely rebuild parts, subassemblies, assemblies, or the entire end item.

    e. <u>Essential Maintenance Action (EMA)</u>: Maintenance that must be performed prior to the next mission. This includes correcting operational mission failures, as well as performing certain unscheduled maintenance actions.

f. Failure: Any single, combination, or summation of hardware or software malfunctions that cause a maintenance action to be performed.

g. Inherent Availability (Ai): Computed with the following formula:

$$A_i = \frac{OT}{OT + TCM}$$

h. Intermediate Level Maintenance (ILM): Maintenance that is authorized by designated maintenance activities in support of using organizations. The principal function of ILM is to repair subassemblies, assemblies, and major items of equipment for return to a lower echelon or to supply channels. Measures are as follows:

(1) MTTR at ILM

(2) MaxTTR at ILM

i. Maintenance Ratio (MR): Computed by the following formula:

$$MR = \frac{\text{Total man-hours of maintenance}}{\text{Operating time}}$$

j. Maximum Time To Repair (MaxTTR): That time below which a specified percentage of all corrective maintenance tasks must be completed.

k. Mean Time Between Operational Mission Failure (MTBOMF): Computed by the following formula:

$$MTBOMF = \frac{\text{Operating time}}{\text{Total number of operational mission failures}}$$

l. Mean Time Between Unscheduled Maintenance Actions (MTBUMA): Computed by the following formula:

$$MTBUMA = \frac{\text{Operating time}}{\text{Total number of unscheduled maintenance actions}}$$

m. Mean Time To Repair (MTTR): Computed by the following formula:

$$MTTR = \frac{\text{Total corrective maintenance time}}{\text{Total number of corrective maintenance actions}}$$

A-3-2

n. <u>Operational Availability (Ao)</u>: Computed by the following formula:

$$A_O = \frac{OT + ST}{OT + ST + TCM + TPM + TALDT}$$

Where:
- OT = Operating time
- ST = Standby time
- TCM = Total corrective maintenance time
- TPM = Total preventive maintenance time
- TALDT = Total administrative and logistics downtime

o. <u>Operating Time (OT)</u>: The period of time that the system is powered and capable of performing a mission-essential function.

p. <u>Operational Mission Failure (OMF)</u>: A failure that reduces the capability of the system to a point where one (or more) mission essential function(s) cannot be performed. Measures are as follows:

    (1)   Mean time between OMF (MTBOMF),

    (2)   Mean miles between OMF (MMBOMF), and

    (3)   Mean rounds between OMF (MRBOMF).

q. <u>Organizational Level Maintenance (OLM)</u>: Maintenance that is authorized for, performed by, and the responsibility of a using organization on its own equipment. Measures are as follows:

    (1)   MTTR at OLM, and

    (2)   MaxTTR at OLM.

r. <u>Preventive Maintenance (PM)</u>: The actions performed to retain an item in a specified condition by systematic inspection, detection, and prevention of incipient failures. The measure is Total Preventive Maintenance (TPM) time.

s. <u>Percent of Correct Detection</u>: Percent of all faults or failures that the BIT system detects.

t. <u>Scheduled Maintenance</u>: Maintenance that is performed on a regular basis over the life of a system in order to maintain its ability to perform mission essential functions. This maintenance consists of programmed services and/or replacements performed at intervals defined by calender time or usage (i.e., miles, hours, rounds...).

A-3-3

u. Standby Time (ST): The period of time that the system
is presumed operationally ready for use, but does not have
power applied, is not being operationally employed, and no PM
or CM is being performed.

v. Time To Repair: A representative time of the effort
expended on corrective maintenance. Measures are as follows:

(1)   Mean time to repair (MTTR), and

(2)   Maximum time to repair (MaxTTR).

w. Unscheduled Maintenance: Maintenance that was not
programmed, but is required to restore the item to partial or
full mission capability.

A-3-4

APPENDIX 4

AIR FORCE TERMS AND DEFINITIONS

1. <u>Purpose</u>. This Appendix provides the RAM terms and definitions which are most relevant to this Annex and used within the Air Force in conducting and reporting OT&E activity. They have been extracted from AFR 55-43, AFP 57-9, and DoD 3235.1-H (Test and Evaluation of System Reliability, Availability, and Maintainability). They are included in the Memorandum of Agreement so as to assist other services in understanding RAM terms as used by the Air Force. Terms used by other services are included in Appendices 1, 2, and 3.

2. <u>Definitions</u>

    a. <u>Break Rate</u>: The percent of time an aircraft will return from an assigned mission with one or more previously working systems or subsystems on the Mission-Essential Subsystem List (MESL) inoperable (code 3 including ground and air aborts). Repairs must be made before the aircraft can perform a subsequent "like-type" mission.

    b. <u>Fix Rate</u>: The percent of aircraft, which return "code 3" from an assigned mission, that must be repaired in a specified number of clock-hours, i.e., 70 percent in 4 hours. Fix rate is similar to mean downtime. The time requirement for fix rate includes direct maintenance time and downtime associated with maintenance policy and administrative and logistics delays.

    c. <u>Maintainability</u>: The ability of an item to be retained in or restored to specified conditions when maintenance is performed by personnel having specified skill levels, using prescribed procedures and resources, at each prescribed level of maintenance and repair.

    d. <u>Maintenance Man-Hours/Operating Hour (MMH/OH)</u>: The number of base-level, direct maintenance man-hours required to support a system divided by the number of operating hours during the period. Where aircraft, ships, and vans are involved, maintenance man-hours/flying hour (MMH/FH), maintenance man-hours/sortie (MMH/S), or some similar requirement may be used.

    e. <u>Maximum Time To Repair (MaxTTR)</u>: The time within which a specified percentage of all corrective maintenance tasks must be completed. For example, 90 percent of all corrective maintenance actions must be completed within two hours.

A-4-1

f. Mean Repair Time (MRT): The average on- or off-equipment corrective maintenance time in an operational environment. MRT starts when the technician arrives at the system or equipment for on-equipment at the system level, and off-equipment at the assembly, subassembly, module, or circuit card assembly at the off-equipment repair location. The time includes all maintenance actions required to correct the malfunction, including preparing for test, troubleshooting, removing and replacing components, repairing, adjusting, and functional check. MRT does not include maintenance or supply delays. MRT is similar to MTTR, but is referred to as MRT when used as an operational term to avoid confusion with the contractual term of MTTR.

g. Mean Downtime (MDT): The average elapsed clock-time between loss of mission-capable status and restoration of the system to mission-capable status. This downtime includes maintenance and supply response, administrative delays, and actual on-equipment repair. In addition to the inherent repair and maintainability characteristics, mean downtime is affected by technical order availability and adequacy, support equipment capability and availability, supply levels, and manning. Thus, MDT is not the same as the contractual term mean time to repair (MTTR).

h. Mean Time Between Critical Failures (MTBCF): The average time between failure of essential system functions. For ground electronic systems, MTBCF is equal to the total equipment operating time in hours divided by the number of failures of essential systems required for completion of the assigned mission. MTBCF includes both hardware and software failures.

i. Mean Time Between Maintenance Events (MTBME): The average time between on-equipment, corrective events including inherent, induced, no-defect, and preventive maintenance actions. It is computed by dividing the total number of life units (for example, operating hours, flight hours, rounds) by the total number of maintenance (base level) events for a specific period of time. A maintenance event is composed of one or more maintenance actions.

j. Mean Time Between Removal (MTBR): A measure of the system reliability parameter related to demand for logistic support. The total number of system life units divided by the total number of items removed from that system during a stated period of time. This term is defined to exclude removals performed to facilitate other maintenance and removals for time compliance technical orders (TCTOs).

k. Mission-Capable (MC) Rate: The percent of possessed time that a weapons system is capable of performing any of its assigned missions. The MC rate is the sum of the full mission-capable (FMC) and partial mission-capable (PMC) rates.

A-4-2

1. Operational Availability: The ability to commit a system or subsystem to operational use when needed, typically stated in terms of A_o, mission-capable rate, sortie generation rate, or uptime ratio. For systems with a defined mission duration, it does not indicate the capability to complete a mission once the mission begins.

m. Percent BIT Can-Not-Duplicate (CND): A BIT CND is an on-equipment, BIT indication of a malfunction that cannot be confirmed by subsequent troubleshooting by maintenance personnel. It is computed with the following formula:

$$\% \ CND = \frac{Number \ of \ BIT \ CND}{Total \ number \ of \ BIT \ indications*} \times 100$$

*Excludes false alarms that do not generate maintenance actions.

n. Percent BIT False Alarm (FA): A BIT FA is an indication of a failure that is not accompanied by system degradation or failure and, in the opinion of the operator, does not require any maintenance action. It is computed by the following formula:

$$\% \ FA = \frac{Number \ of \ BIT \ indications \ not \ resulting \ in \ maintenance \ actions}{Total \ number \ of \ BIT \ indications} \times 100$$

o. Percent BIT Fault Detection (FD): Measures instances where a confirmed failure is a condition when equipment performance (including BIT performance) is less than that required to perform a satisfactory mission, and corrective action is required to restore equipment performance. The formula below assumes that a requirement exists for 100-percent diagnostics capability.

$$\% \ BIT \ FD = \frac{Number \ of \ confirmed \ failures \ detected \ by \ BIT}{Number \ of \ confirmed \ failures \ detected \ via \ all \ methods} \times 100$$

p. Percent Fault Isolation (FI): It is just as opera-tionally valuable for BIT to fault-isolate an aircrew-reported fault, or manually detected fault, as it is for BIT to fault-isolate BIT-detected faults. Effective isolation means that the fault is unambiguously isolated to a single-item node (driver, receiver, connector, wire) or to a specified maximum

A-4-3

number of items (an ambuiguity group of x items). The formula below defines the percent of FI.

$$\% \text{ FI} = \frac{\text{Number of fault isolations in which BIT effectively contributed}}{\text{Number of confirmed failures detected via all methods}} \times 100$$

q. <u>Percent Retest OK (RTOK)</u>: Defined by the formula below as follows:

$$\% \text{ RTOK} = \frac{\text{Number of units (LRU, SRU) that RTOK at a higher maintenance level}}{\text{Number of units removed as a result of BIT}} \times 100$$

r. <u>Uptime Ratio (UTR)</u>: The percentage of time that operational equipment is able to satisfy mission demands. UTR is similar to MC, except that system status depends on current use of the system, as well as the designated operational capability (DOC). For example, a system with several DOC missions can be MC if at least one of those missions can be accomplished. However, if an immediate need exists for a mission capability that is "down", the overall system is considered to be "down."

s. <u>Weapon System Reliability (WSR)</u>: The probability that a system will complete a specified mission given that the system was initially capable of doing so.

BIBLIOGRAPHY

GENERAL TEXT AND REFERENCE BOOKS

Blanchard, B. S. *Logistics and Engineering Management*. Englewood Cliffs, N.J.: Prentice-Hall, Inc., 1974.

Blanchard, B. S. and E. E. Lowery. *Maintainability Principles and Practices*. New York: McGraw-Hill Book Co., Inc., 1969.

Brownlee, K. A. *Statistical Theory and Methodology In Science and Engineering*. New York: John Wiley and Sons, Inc., 1965.

Comfort, G. C., et. al. *Examination of DT&E Planning, Volume I--DT&E Planning Indicators: Linking Development Test and Evaluation to Operational Performance*, Alexandria, Va.: Institute for Defense Analyses, IDA Report R-323, December 1987.

Cramer, H. *Mathematical Methods of Statistics*. Princeton, N.J.: Princeton University Press, 1946.

Feller, W. *An Introduction to Probability Theory and Its Applications. Vol. I*. New York: John Wiley and Sons, Inc., 1957.

Feller, W. *An Introduction to Probability Theory and Its Applications. Vol. II*. New York: John Wiley and Sons, Inc., 1966.

Mann, G. A. *The Role of Simulation in Operational Test and Evaluation*. Maxwell AFB, Ala.: Airpower Research Institute, August 1983.

Mood, A. M. and F. A. Graybill. *Introduction to the Theory of Statistics*. New York: McGraw-Hill Book Co., Inc., 1963.

Morgan, C. T., et al. *Human Engineering Guide to Equipment Design*. New York: McGraw-Hill Book Co., Inc., 1963.

Shigley, J. E. *Simulation of Mechanical Systems; An Introduction*. New York: McGraw-Hill Book Co., Inc., 1967.

Spiegel, M. R. *Theory and Problems of Statistics*. New York: McGraw-Hill Book Co., Inc., 1961.

Stevens, R. T. *Operational Test and Evaluation: A Systems Engineering Process*. New York: John Wiley & Sons, Inc., 1979

Uspensky, J. V. *Introduction to Mathematical Probability*. New York: McGraw-Hill Book Co., Inc., 1937.

Waugh, A. E. *Elements of Statistical Method*. New York:McGraw-Hill Book Co., Inc., 1938.

GOVERNMENT DOCUMENTS

_____. *Acquisition of Major Defense Systems*. DOD Directive 5000.1, July 13, 1971.

_____. *Briefing for the OTA Commanders' Conference, Common Terminology Working Group, Progress Report*, August 11, 1987.

169

_____. *Comments on DOD's Implementation of Recent Procurement Reforms*, GAO Document T-NSIAD-87-23, March 31, 1987.

_____. *Defense Acquisition Program Procedures*, DoD Instruction 5000.2, September 1, 1987.

_____. *Director of Operational Test and Evaluation*, DoD Directive 5141.2, April 2, 1984.

_____. *DOD's Defense Acquisition Improvement Program: A Status Report*, GAO Report 86-148, July 23, 1986.

_____. *Force Development, Operational Test and Evaluation, Methodology and Procedures Guide*, Department of the Army Pamphlet 71-3, Draft, March 1988.

_____. *Logistics Assessment*, AFOTEC Pamphlet 400-1, February 29, 1984. With change 1, March 31, 1986.

_____. *Management of Software Operational Test and Evaluation*, AFOTEC Pamphlet 800-2, Volume 1, August 1, 1986.

_____. *Operational Test and Evaluation Can Contribute More to Decisionmaking*, GAO Report NSIAD-87-57, December 23, 1986.

_____. *Operational Test Director Guide*, Department of the Navy, Commander Operational Test and Evaluation Force, COMOPTEVFOR Instruction 3960.1E, June 27, 1986, with change 2 dated March 2, 1987.

_____. *Reliability Tests: Exponential Distribution*. MIL-STD-781B. U.S. Government.

_____. *Research, Development, and Acquisition Procedures*, Department of the Navy, OPNAV Instruction 5000.42C, May 10, 1986.

_____. *Software Maintainability Evaluation Guide*, AFOTEC Pamphlet 800- 2, Volume 3, January 28, 1988.

_____. *Software Management Indicators*, Air Force Systems Command, AFSCP 800-43, January 31, 1986.

_____. *Software Quality Indicators*, Air Force Systems Command, AFSCP 800-14, January 20, 1987.

_____. *Software Support Life Cycle Process Evaluation Guide*, AFOTEC Pamphlet 800-2, Volume 2, Draft, March 1, 1988.

_____. *Software Support Resources Evaluation Guide*, AFOTEC Pamphlet 800-2, Volume 5, May 2, 1988.

_____. *Software Test and Evaluation Manual*, DoD 5000.3-M-3, November 1987.

_____. *Software Usability Evaluator's Guide*, AFOTEC Pamphlet 800-2, Volume 4, November 23, 1987.

_____. *Status of Space Surveillance and Tracking System*, GAO Report NSIAD 88-61, November 10, 1987.

_____. *Systems Engineering Management.* MIL-STD-499 (USAF). Air Force Systems Command, July 1969.

_____. *Test and Evaluation.* DOD Directive 5000.3, March 12, 1986.

_____. *Test and Evaluation,* Draft DoD Directive 5000.3, August 1988 Proposed Revision.

_____. *Test and Evaluation,* Department of the Navy, OPNAV Instruction 3960.10C, September 14, 1987.

_____. *Test and Evaluation Master Plan (TEMP) Guidelines,* DoD 5000.3- M-1, October 1986.

_____. *Test and Evaluation of Navy Systems and Equipment.* OPNAV Instruction 3960.8, January 22, 1973.

_____. *Test and Evaluation of System Reliability, Availability, and Maintainability,* DoD Instruction 3235.1, February 1, 1982.

_____. *Test & Evaluation of System Reliability, Availability, and Maintainability--A Primer,* DoD 3235.1-H, July 1981.

_____. *Testing Oversight, Operational Test and Evaluation Oversight: Improving but More is Needed,* GAO Report NSIAD-87-108BR, March 18, 1987.

_____. *The Advanced Tactical Fighter's Costs, Schedule, and Performance Goals,* GAO Report NSIAD 88-76, January 13, 1988.

_____. *USAF R&M 2000 Process,* Headquarters USAF, October 1987.

_____. *Weapons Testing, Quality of DOD Operational Testing and Reporting,* GAO Report PEMD-88-32BR, July 26, 1988.

ADM	Acquistion Decision Memorandum
AFOTEC	Air Force Operational Test and Evaluation Center
ALDT	Administrative Logistics Delay Time
ATE	Automatic Test Equipment
AVSCOM	Aviation System Command
BCM	Baseline Correlation Matrix
BIT	Built-In Test
BITE	Built-In Test Equipment
BIT/FIT	Built-In Test and Fault Isolation Test
CBR	Chemical, Biological, Radiation
CFE	Contractor-Furnished Equipment
CND	Cannot Duplicate
COI	Critical Operational Issue
DAB	Defense Acquisition Board
DAG	Data Authentication Group
DCP	Decision Coordinating Paper
DEM/VAL	Demonstration/Validation Phase
DoD	Department of Defense
DOT&E	Director, Operational Test and Evaluation
DSM	Data Source Matrix
DT	Developmental Test
DT&E	Developmental Test and Evaluation
E3	Electromagnetic Environmental Effects
EDM	Engineering Development Models
EMA	Essential Maintence Action
EMC	Electromagnetic Compatability
EMI	Electromagnetic Interference
FA	False Alarm
FD	Fault Detection
FI	Fault Isolation
FMC	Full Mission-Capable
FOC	Full Operational Capability
GAO	General Accounting Office
GPS UE	Global Positioning Satellite User Equipment
HELO	Helicopter
IER	Independent Evaluation Report
ILS	Integrated Logistics Support
ILSMT	Integrated Logistics Support Management Team
ILSP	Integrated Logistics Support Plan
IOC	Initial Operational Capability
IOT&E	Initial Operational Test and Evaluation
LRIP	Low Rate Initial Production
LRU	Line Replaceable Unit
LSA	Logistics Support Analysis
MANPRINT	Manpower and Personnel Integration
MaxTTR	Maximum Time To Repair
MC	Mission-Capable
MCMT	Mean Corrective Maintenance Time
MCMT	Mean Corrective Maintenance Time
MCOTEA	Marine Corps Operational Test Activity
MESL	Mission-Essential Subsystem List

MIL-STD	Military Standard
MMBUMA	Mean Miles Between Unscheduled Maintenance Actions
MMH/OH	Maintenance Man-Hours required per Hour of Operation
MNS	Mission Need Statement
MOA	Memorandum of Agreement
MOE	Measures of Effectiveness
MOMFRT	Mean Operational Mission Failure Repair Time
MOT&E	Modification Operational Test and Evaluation
MPs	Military Police
MR	Maintenance Ratio
MSI	Maintenance Support Index
MTBEMA	Mean Time Between Essential Maintenance Actions
MTBF	Mean Time Between Failure
MTBMA	Mean Time Between Maintenance Actions
MTBOMF	Mean Time Between Operational Mission Failures
MTBR	Mean Time Between Removal
MTBUM	Mean Time Between Unscheduled Maintenance
MTTFL	Mean Time To Fault Locate
MTTR	Mean Time to Repair
MTTRF	Mean Time To Restore Function
MWS	Missile Warning System
M&S	Modeling and Simulation
NDI	Non-Developmental Items
OASD(P&L/WSIG)	Office of the Assistant Secretary of Defense (Production and Logistics/Weapons Support Improvement Group)
OMF	Operational Mission Failure
OPTEVFOR	Operational Test and Evaluation Force
OSD	Office of the Secretary of Defense
OTA	Operational Test Agency
OTEA	Operational Test and Evaluation Agency
OT&E	Operational Test and Evaluation
PE	Program Element
PM	Program Manager
PMC	Partial Mission Capable
P(cd)	Percent of Correct Detection
RAM	Reliability, Availability, Maintainability
ROC	Required Operational Capability Statement
RPV	Remotely Piloted Vehicle
SA	Staff Assistant
SARs	Selected Acquisition Reports
SCP	System Concept Paper
SON	Statement of Operational Need
SOP	Standard Operating Procedure
SROs	System Readiness Objectives
SRU	Shop Replaceable Unit
TD	Test Director
TECOM	Test and Evaluation Command
TEMP	Test and Evaluation Master Plan
TMDE	Test Measurement Diagnostic Equipment
TPWG	Test Plan Working Groups
TR	Test Report
TRADOC	Training and Doctrine Command
TWIG	Test Integration Working Group
T.O.s	Technical Orders
USALEA	U. S. Army Logistics Evaluation Agency

SUBJECT INDEX

176

179